BOTTLING, PICKLING & PRESERVING

Mary Makarem
April, in London.

BOTTLING, PICKLING & PRESERVING

ELISABETH LAMBERT ORTIZ

DK
DORLING KINDERSLEY
NDON • NEW YORK • SYDNEY • MOSCO

A DORLING KINDERSLEY BOOK

Created and produced by
CARROLL & BROWN LIMITED
5 Lonsdale Road
London NW6 6RA

Editorial Director Jeni Wright
Editors Julia Alcock
Angela Nilsen

Art Editor Chrissie Lloyd
Designers Carmel O'Neill
Michael Dyer
Production Wendy Rogers
Amanda Mackie

First published in Great Britain in 1994 by Dorling Kindersley Limited,
9 Henrietta Street, London WC2E 8PS

First published as a Dorling Kindersley paperback 1998

Originally published as Clearly Delicious

Visit us on the World Wide Web at http://www.dk.com

A CIP catalogue record for this book is available from the
British Library.

ISBN 0 7513 0645 2

Reproduced by Colourscan, Singapore
Printed and bound in Singapore by Star Standard Industries (Pte.) Ltd.

Foreword

The art of preserving can turn an ordinary storecupboard or pantry into an Aladdin's cave of good things – beautiful to look at, luscious to eat, and a joy both to give and receive. It can also turn your refrigerator and freezer into a rescue service for emergiences like anniversaries – when presents are obligatory, but unbought. The kitchen is a cheerful place, and although, undeniably, there is work involved in transforming fruits into jams and jellies, and vegetables into pickles and chutneys, there is no need for lonely drudgery when it comes to preserving. Family, and often friends, like to join in, and what started as a kitchen chore can turn into a celebration with everyone, not just the cook, having fun.

Being creative and economical at the same time can bring a glow of virtue to even the most modest cheek. But sometimes imagination flags and the cook is bereft of ideas for a meal that is looming. An array of preserves in the storecupboard, refrigerator or freezer will stimulate culinary wits, and new dishes may well be created by the challenge these good things pose: new flavours to liven up dull ordinary dishes or simply help familiar ones that need a lift.

The conservation-minded can take special pleasure from re-using jars, bottles, tins, and boxes that would otherwise clutter up kitchen shelves, reproachfully useless. Those artistically gifted with clever fingers can do wonders with gift wrapping, decorating glass jars and bottles, and transforming tins and baskets with ingenuity but little cost. Wrapping paper, labels and tags, ribbons, and cord can all be recruited to help the container be worthy of its contents.

As the Spanish saying has it: "*es mas bueno el vino en bella copa*" – the wine is better in a beautiful glass.

Elisabeth Lambert Ortiz

Contents

Foreword 5

Introduction 8-9

Beautiful Bottles 10
Sterilizing Bottles, Jars & Closures 11

Jams, Jellies & Other Sweet Concoctions 12-53

Imaginative fruit and sugar combinations

Jams 14
Equipment 16
Recipes 18-27
Preserves & Conserves 28
Recipes 29-31
Marmalades 32
Recipes 33-39
Jellies 40
Equipment 40
Recipes 41-46
Fruit Butters & Cheeses 47
Recipes 48-50
English Fruit Curds 51
Recipes 52-53

Fruits in Alcohol & Flavoured Drinks 54-69

Fruits and spices take on delicious flavours from a variety of alcohols, and vice versa

Fruits in Alcohol 56
Recipes 57-61
Flavoured Wines & Spirits 62
Recipes 63-65
Fruit Cordials & Syrups 66
Recipes 67-69

Pickles, Chutneys, Relishes & Mustards 70-105

Perfect accompaniments to savoury dishes – these recipes are mostly vinegar-based

Vegetable & Fruit Pickles 72
Equipment 74-75
Recipes 76-89
Chutneys 90
Recipes 91-99
Relishes 100
Recipes 101-103
Mustards 104
Recipes 105

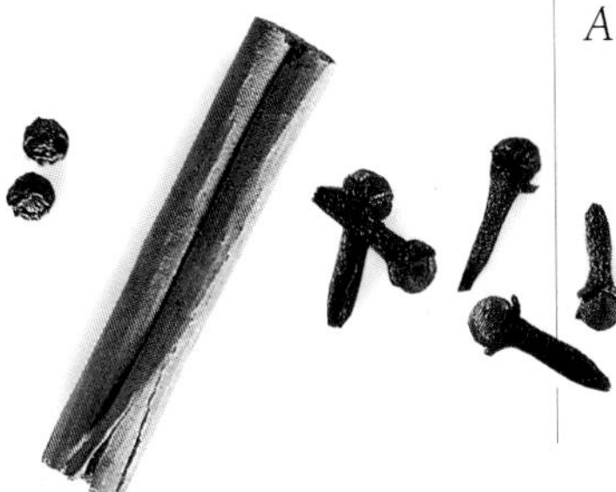

Luxurious Oils & Vinegars 106-121

Oils and vinegars emulsify, flavour, and preserve

Flavoured Oils 108
Recipes 109-112
Preserves in Oil 113
Recipes 114-116
Flavoured Vinegars 117
Recipes 118-121

Herbs & Spices 122-131

A host of seasonings for savoury recipes, sweet dishes, and alcohols

Herb & Spice Blends 124
Recipes 125-131

Finishing Touches 132-140

Advice on how to decorate your produce, for your own pleasure and for gift-giving

Decorating Bottles 134
Decorating Jars 136
Presentation Ideas 138
Labels & Tags 140

Index 141-143

Acknowledgments 144

Introduction

The house becomes filled with utterly delectable perfumes when the seasons bring round the time for making preserves.

The practice of "putting up" summer surpluses for winter eating was born of necessity in the days before refrigeration and worldwide imports of fresh fruits and vegetables.

What was then a necessity has today become a pleasure. What can you do with all those tomatoes that ripen at once, or a tree full of fruits, ready for picking before the birds beat you to it, or green beans that will grow in days from delicate little things to tough old monsters?

Transform them by preserving them, enjoy them until the next harvest, and generously give some away. There are so many occasions for parting with a jar from your storecupboard, refrigerator, or freezer, that it is an excellent idea to put up a lot of good things so as not to run out and find the pantry bare – and with spring an age away. Be as lavish as nature is, a model of generosity. The techniques used in preserving are not difficult, and most kitchens have all the equipment that is needed. Home preserving is a great help to those seeking to avoid additives and preservatives in their food. It is also a great help to the economy-minded, who want luxury foods at a reasonable price, and for the value-conscious, who know they can create specialities at home for a fraction of the cost of the store-bought article. The key to successful preserving (in addition to being a creative cook) is always to use the best and freshest ingredients – easy, since both fruits and vegetables are cheapest when they are at their seasonal best, at the most lavish time of harvest.

In these conservation-minded times, we have the added bonus of beautifully designed, modestly priced pressed-glass jars in which to put our preserves and, in addition, equally well-designed jars and bottles that have housed our everyday store-bought foods – olives, mustards, capers, spices, and oils, to name but a few. Too pretty and elegantly shaped to throw away, they accumulate on pantry shelves until, washed and with labels removed, they stand ready to be filled with all manner of good things, making us good conservationists as well as good cooks.

But remember, re-used bottles and jars without proper seals are suitable for short-term storage only, and are best kept in the refrigerator or freezer. For long-term storage, choose proper preserving containers and lids, and decant your preserves into beautiful bottles to bring to the table or to give as gifts, telling the recipient to store them correctly or eat up quickly!

The occasions for gift-giving are almost endless. There are housewarming gifts, a gift to a new neighbour, gifts to take to a Christening party, a thankyou gift, or one to take to a lunch or dinner party instead of chocolates, fresh flowers, or wine. In fact, a preserve is just the thing to take to any special occasion or gathering.

The pleasure of giving and receiving gifts is as old as history, enshrined in legend and tradition, and kept alive principally because it is such a joy. It is a great idea to revive the old-fashioned way of making preserves for gift-giving; transforming fruits and vegetables into pickles, conserves, jams, jellies, curds, marmalades, syrups, flavoured vinegars, fruits in alcohol, relishes, chutneys, and more. In addition, the new twist of having modern tools like the food processor, and equipment like freezers, lightens your workload. We now have all the world's ingredients to choose from, starting with the herbs grown in our own gardens, then farms where you can pick your own fresh fruits, to summer and autumn wayside farm stands, and nearer to home our very own greengrocers, speciality food shops, delicatessens, and supermarkets.

Putting up the bounty of summer can take the hassle out of Christmas, especially for those who hate shopping. There is no need for the punishing business of last-minute forays into stores, or the awful embarrassment of having forgotten someone when you don't have time to rush out and buy a gift. The holidays are a wonderful time to part with luxury preserves, such as Seville marmalade with almonds, pawpaws bottled with rum and pistachios, redcurrant jelly with port, or a special liqueur flavoured with fruits picked from your own garden.

It is important to label jars and bottles, because it is surprisingly easy to forget which contents are which. The date should always be added, because it is equally easy to forget when they were made. A simple factual label is fine for preserves in the family storecupboard, refrigerator, or freezer, but for gifts, buy or make labels that will turn the jar or bottle into a decorative container. Anyone clever with their hands can use a felt-tip or calligraphy pen to write fancy lettering for the labels, while small swatches of patterned fabric tied over the tops of jars with coloured ribbon transform even the plainest glass jar into something special. Jams, jellies, and preserved whole fruits are often attractive enough in themselves that they need no further embellishment. However, decorating your containers is fun, and a rewarding finishing touch to the fruits of your labour.

BEAUTIFUL BOTTLES

The rich colours and enticing appearance of preserves are best exploited by displaying them in glass containers. Glass is an excellent material, because it does not react with any of the ingredients used, and it tolerates careful heating. All kinds of attractive and unusual shapes are available and any type of reasonably thick glass bottle or jar can be used, as well as specialized preserving jars, which are more suitable for long-term storage. Whichever containers you choose, they must be scrupulously clean before use. The size and shape of bottles and jars are often dictated by the type of preserve you are making. Large jars with wide necks are needed for packing whole fruits into, whereas 500 ml jars are more useful for jams, jellies, and chutneys.

Sterilizing Bottles, Jars, & Closures

Before you start making any preserve, it is important that all bottles, jars, and closures be clean, free of cracks, and sterile. Here is a quick step-by-step guide to the sterilizing technique.

- Clean off all labels (if re-using containers) and wash the bottles, jars, and closures in hot water with detergent, then rinse thoroughly in hot water.
- Put a wire rack in the bottom of a large deep saucepan, and place the bottles and jars on it.
- Pour in enough hot water to cover the bottles and jars completely. Bring to a boil and boil rapidly for 10 minutes. Remove from the water and turn them upside-down on a thick tea-towel to drain.
- Sterilize the closures by dipping them in boiling water. Leave the closures to dry thoroughly on a tea-towel.
- Preheat the oven to 110°C (225°F). Dry the sterilized bottles and jars in the oven. They can be kept warm in the oven until required or, if the preserve is to be put into a cold container, remove them, leave to cool, and fill as soon as possible.

A Word of Caution

Re-used bottles and jars are suitable for short-term storage, or for storage in the refrigerator or freezer, but for long-term keeping in the storecupboard, be sure to use proper preserving jars and bottles with airtight closures and seals. For gift-giving and taking to the table, you can always decant your preserves into more attractive new or recycled bottles or jars.

Stand the bottles and jars, right-side up, on a rack in the pan, making sure that they do not touch either the sides of the pan or each other.

All closures should be scalded in boiling water before using. Lower the closures into the boiling water with a pair of tongs to eliminate the risk of burning hands.

Dry the sterilized jars, right-side up, on a baking sheet, in a preheated low oven. This will take about 15 minutes. Do not increase the heat, or the glass may crack under pressure.

Bottles on Parade

Choose a shape and size of bottle or jar to complement the preserve. Plain glass is ideal for clear, shimmering mixtures, such as jellies, while subtly patterned ones work well with flavoured oils and vinegars.

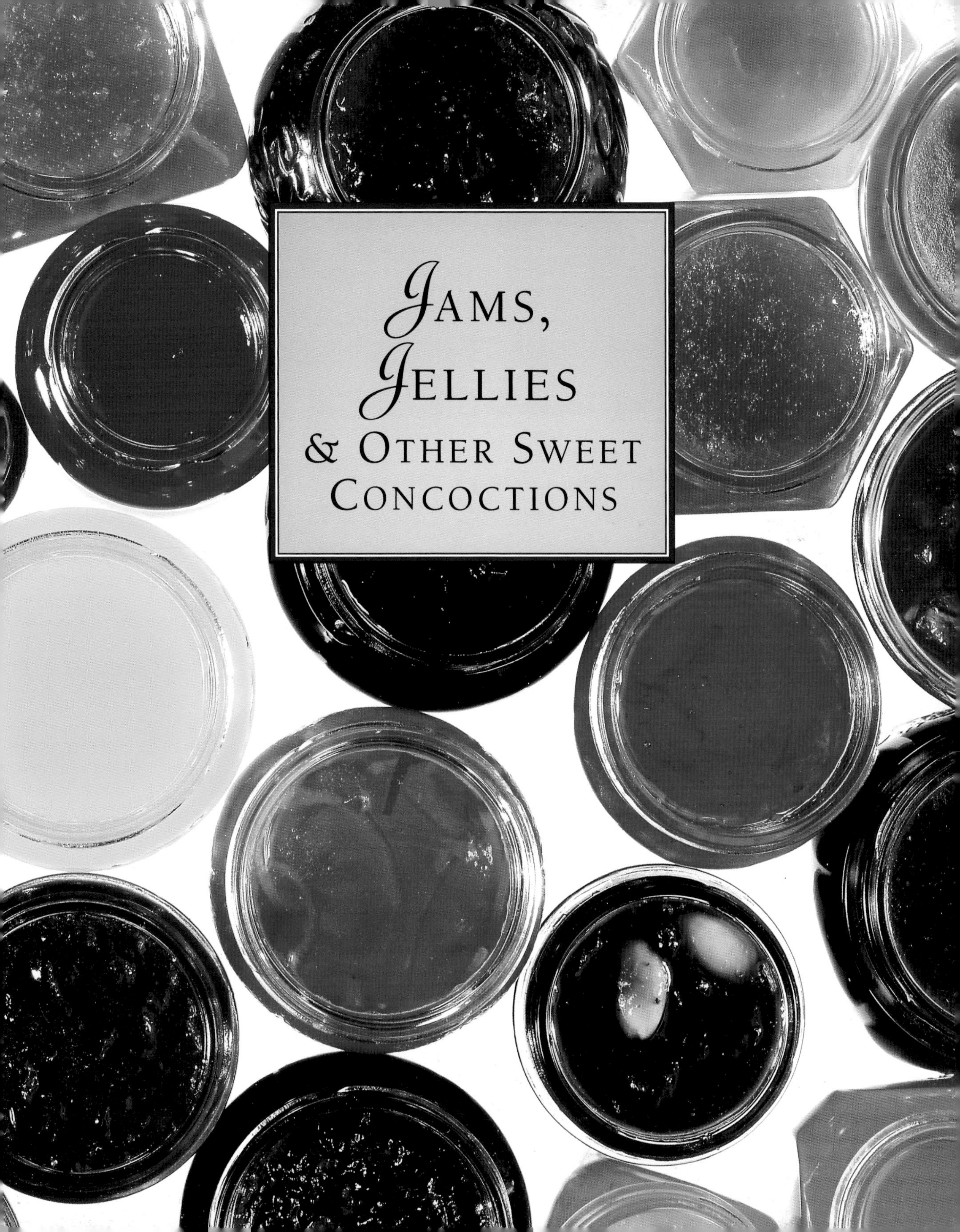

Jams, Jellies & Other Sweet Concoctions

JAMS

SUMMER HERALDS the joyous beginning of jam making, when the house is filled with the perfume of simmering fruits being transformed by sugar and heat into a luscious spread. Jam turns bread into a luxury, and is a useful ingredient in sauces, puddings, and desserts, as well as in cake and biscuit making. There is hardly a sight more enticing than an abundance of fruits waiting to be transformed into jam. With a great cornucopia of fruits to choose from, swelling and ripening in a tantalizing succession of harvests, throughout summer and autumn, there is plenty of room for experimentation.

Although familiar fruits, such as raspberries and blackcurrants, still provide the most popular jams, less traditional flavours can be created with pineapples, or apples and ginger. Indeed, part of the fun when filling jars with the finished jam, is to admire the colours produced by the clever blending of fruits. Rhubarb and strawberries, for instance, combine to make a beautiful rose-red confection, pleasing to both eye and palate.

MAKING JAMS

❦ Choose the ingredients. Jam is made with two main ingredients – fruit and sugar. The fruits must contain the right proportions of pectin and acid so that the jam can set properly, and enough sugar must be added to preserve it. (Pectin is a natural strengthening substance found only in fruits). Fruits that are fresh and slightly under-ripe contain the most pectin, so buy or pick the desired fruits when they are just coming into season. Over-ripe fruits contain very little pectin so they do not set well, and are better kept for making chutneys. Fruits high in both pectin and acid are apples, cranberries, gooseberries, currants, and tart red plums. Fruits that will set a jam moderately well are blackberries, apricots, loganberries, raspberries, and sweet green plums. Cherries, pears, pineapples, nectarines, and some varieties of strawberry are poor setters and need the addition of pectin and acid to make a satisfactory jam. Fruits that do not have good setting properties can be used in combination with those that do, such as pears with tart red plums, and raspberries with redcurrants; these combinations may also require the addition of an acid, such as lemon juice. Alternatively, commercial fruit pectin can be useful for adding to jams made with ingredients low in pectin. It is important to add the correct amount, so follow the manufacturer's instructions carefully. Liquid pectin is used in these recipes; if unavailable, use in powdered form. Dried fruits like apricots make quite good jam, provided they have not been preserved with sulphur dioxide, which can affect the set. They need to be soaked for at least 24 hours before cooking. Some vegetables, such as marrows and carrots, can be used to make jam, but they need plenty of added lemon juice to ensure a reasonable set, and may require extra flavourings, such as spices.

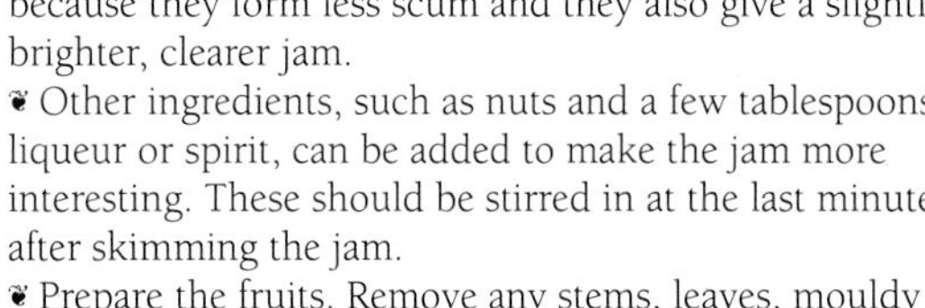

❦ When it comes to the choice of sugar, both refined cane and beet granulated sugars work well, and there is no difference in the keeping qualities ot jam made from either of these. However, lump and coarse sugar are better because they form less scum and they also give a slightly brighter, clearer jam.

❦ Other ingredients, such as nuts and a few tablespoons of liqueur or spirit, can be added to make the jam more interesting. These should be stirred in at the last minute after skimming the jam.

❦ Prepare the fruits. Remove any stems, leaves, mouldy or damaged parts. Wash lightly and pat dry with kitchen paper, to remove excess water. Stone fruits should be halved and stoned. Some recipes call for the addition of a few kernels and these can be extracted by cracking the stones with a hammer. Adding a few kernels will give the

jam extra flavour, but too many will make the jam bitter. Blanch the kernels in boiling water and remove their skins before adding to the fruit.

❦ Weigh out and measure the ingredients carefully, because accuracy is important for successful setting.

❦ Put the prepared fruits into a preserving pan with the recommended amount of water, and simmer until the fruits are soft. Generally, the mixture should reduce by about one-third before the sugar is added to ensure a good set. Start the juices running in soft fruits by crushing some fruits in the bottom of the pan.

❦ Warm the sugar (see box, page 20) while the fruits are cooking. Although this is not essential, the sugar will dissolve more quickly when warm rather than when cold. Make sure the fruit is really soft; if the sugar is added too soon, it will have a hardening effect on the fruits and this cannot be rectified. Add the sugar all at once and stir over a low heat until it has completely dissolved. A knob of butter can be added at this point to help reduce foaming, but this is not necessary. Increase the heat and bring to a boil, then simmer, stirring occasionally, until setting point is reached. Stir the jam as little as possible after the sugar has dissolved.

❦ Test the jam for set, removing it from the heat first to prevent over-cooking. The best way to test is with a sugar thermometer. If you can, clip the thermometer to the side of the pan when you start to cook the fruits. You will then be able to read the temperature as the jam progresses. If this is not possible, heat the thermometer in a jug of hot water. Stir the jam and immerse the bulb of the thermometer completely in the jam, taking care not to let it touch the bottom of the pan. When the thermometer reads 104°C (220°F), the jam should set if you have followed the recipe carefully.

❦ You can also test the jam by dropping a little on to a cold plate. Chill quickly in the refrigerator. If the jam forms a skin, and wrinkles when it is pushed with a finger or spoon, it should have reached setting point.

❦ When setting point is reached, lightly skim off any scum with a long-handled metal spoon. Pour the jam into warmed, dry sterilized jars (see page 11). Fill the jars to within 3 mm of the tops. If directed in the recipe, leave whole fruit jams to cool slightly before putting into jars, so that a thin skin forms and the fruits are evenly distributed through the jam.

How to seal and store

Seal the jars tightly, either with screw-top lids with plastic linings, or with ordinary metal or hard plastic twist-off lids. Jams can also be sealed with wax and cellophane circles. For this method, place the correct size of wax circle , wax-side down, on the surface of the jam. It should cover the jam exactly, and not run up the sides of the jar. Smooth over with a fingertip to remove any air pockets that may be trapped between the jam and the wax circle. Cover with a dampened cellophane circle and secure firmly with a rubber band. Label the jars and keep in a cool dark place or in the refrigerator. Paper covers do not give such an airtight seal, so they are best for short-term storage only.

What can go wrong and why

A jam will crystallize if an over-concentration of sugar has been caused by excessive boiling. Jam can ferment or go mouldy if it contains too little sugar, or if it was not boiled sufficiently. This also happens in poor storage conditions, or through using over-ripe, wet, or bad fruits, or wet jars. If the jam is very dark, it has probably been boiled too fast at the first stage or too slow at the sugar stage. Storage in too bright a light can also darken the jam. If setting point has not been reached, the jam will be runny. This can be corrected by re-boiling the runny jam. Alternatively, add commercial fruit pectin and follow the instructions on the package carefully (too much pectin will spoil the flavour of the jam). Other setting problems that occur with jams can be caused by a lack of pectin or acid, or by under-boiling so that the pectin is not fully released.

EQUIPMENT

Jam making requires little in the way of specialist equipment. The most worthwhile investment for the avid jam maker is a good-quality preserving pan. Most of the other items shown here are pieces of everyday equipment.

CITRUS FRUIT SQUEEZER

STAINLESS STEEL SPOON
Plain or slotted for skimming.

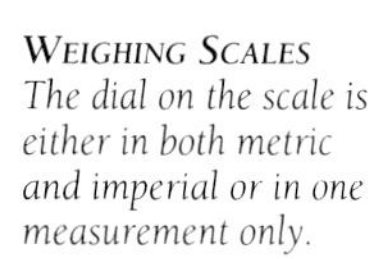

MEASURING JUG
Plastic, china, glass, or stainless steel measuring jugs are all suitable. Other metals should be avoided, particularly if lemon juice is being used.

WEIGHING SCALES
The dial on the scale is either in both metric and imperial or in one measurement only.

LADLE

SUGAR THERMOMETER

CHOPPING BOARD, SHARP STAINLESS STEEL KNIVES, AND VEGETABLE PEELER

MEASURING SPOONS

MUSLIN
For wrapping up spices or pith and pips, and for straining.

WOODEN SPOONS
Choose long-handled ones so that the hot preserve doesn't splash your hand.

DID YOU KNOW?
A wide variety of jam jars and closures are readily available. The most effective container for long-term storage is a glass jar with a screw-top lid, because the lid is airtight. Wax and cellophane circles are also very useful for covering jams on a short-term basis. The circle of wax is placed on the jam, wax-side down, and smoothed over carefully with a fingertip to remove any air pockets that may be trapped between the jam and the wax circle. A damp cellophane circle is then placed over the jar and firmly secured with a rubber band.

PLASTIC SIEVES
Metal sieves can give an unpleasant flavour to jam, so always use plastic ones.

LABELS

PRESERVING PAN
A wide, fairly shallow preserving pan, made of good-quality stainless steel, is best for rapid boiling of fruit and sugar. Avoid brass or copper, unless specified in the recipe.

CONTAINERS
Use sterilized and dried jars and closures.

FUNNELS
Essential for pouring liquids from one container to another, especially bottles with narrow necks.

PLUM AND WALNUT JAM

Late summer is the time to make this jam, and plums, in all their different hues, offer a wide choice of colours. Try blending red and yellow plums to give a deep shade of pink to your pots of jam. Chopped nuts stirred in at the last minute make an interesting addition.

INGREDIENTS

1 kg red plums

1 kg yellow plums

450 ml water

1.8 kg sugar, warmed (see box, page 20)

225 g walnuts, roughly chopped

Makes about 2.75 kg

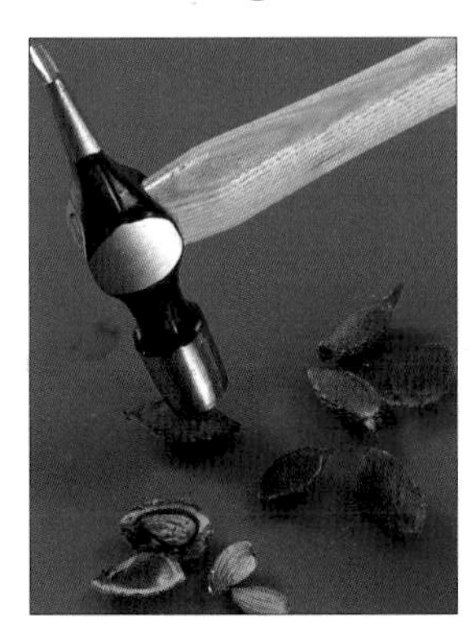

1 ◂ Halve and stone the plums. Crack a few of the stones with a hammer and take out the kernels. Discard the rest.

2 Put the kernels into a bowl and pour over boiling water to cover. Leave for 1 minute, then drain and transfer to a bowl of cold water. Drain again, and rub off the skins.

3 ◂ Put the plums, kernels, and measured water into a preserving pan and bring to a boil. Lower the heat and simmer, stirring occasionally, for 30–40 minutes, or until the plum skins are soft and the fruit is tender. The mixture in the pan should have reduced by about one-third.

4 ◀ Add the warmed sugar, and stir over a low heat, until the sugar has completely dissolved. Increase the heat and boil the mixture rapidly, without stirring, for 10 minutes, or until it reaches setting point. Remove the pan from the heat to test.

5 ▼ The sugar thermometer should read 104°C (220°F). If you do not have a thermometer, test for set with the cold plate test (see page 15). Lightly skim off any scum, using a long-handled metal spoon.

6 ▲ Stir in the walnuts. Immediately pour the jam into warmed sterilized jars, to within 3 mm of the tops. Seal the jars and label.

LATTICED PLUM PIE
Plum and Walnut Jam fills this luscious tart, topped with a decorative lattice of pastry.

PEACH JAM

INGREDIENTS

2 kg peaches

300 ml water

1.8 kg sugar, warmed (see box, below)

75 ml lemon juice

120 g package liquid fruit pectin

Most liquid fruit pectin is made from the white pulp under the skin of citrus fruit, but it can also be made from apples.

1 Halve, stone, peel, and dice the peaches. Crack a few of the stones with a hammer and take out the kernels. Discard the rest. Put the kernels into a small bowl and pour boiling water over to cover. Leave for 1 minute, then drain and transfer to a bowl of cold water. Drain again, then rub off the skins with your fingers.

2 Put the peaches, kernels, and the measured water into a preserving pan and bring to a boil. Lower the heat and simmer, stirring occasionally, for 20–30 minutes, or until the peaches are soft. Add the warmed sugar, lemon juice, and pectin, and stir with a wooden spoon, until the sugar has completely dissolved.

3 Increase the heat and boil the mixture rapidly, without stirring, for only 1 minute or according to package instructions. With the pan off the heat, lightly skim off any scum from the surface of the jam, using a long-handled metal spoon. Cool the jam slightly.

4 Pour the jam into warmed sterilized jars, to within 3 mm of the tops. Seal the jars and label.

Makes about 2.25 kg

DID YOU KNOW?
Jam will not set without pectin, which is a setting agent found naturally in most fruits, in varying degrees. Recipes made with fruits that are low in pectin are best made with a commerically bought liquid pectin to ensure a good set.

PEAR AND PLUM JAM

INGREDIENTS

1 kg Conference pears

1 kg tart red plums

300 ml water

1.8 kg sugar, warmed (see box, below)

1 Halve, core, peel, and dice the pears. Put the peel and cores on a square of muslin and tie up tightly into a bag with a long piece of string. Put the diced pears, plums, and water into a preserving pan. Tie the muslin bag to the pan handle, so that it rests on the fruit, and bring to a boil.

2 Lower the heat and simmer, stirring occasionally, for at least 1 hour, or until the plum skins are soft and the fruit is tender. The mixture in the pan should have reduced by about one-third. Skim off the plum stones with a slotted spoon as they come to the surface, and discard. Discard the bag, squeezing it first to extract all the juice.

3 Add the warmed sugar to the fruit mixture, and stir over a low heat, until the sugar has completely dissolved.

4 Increase the heat and boil the mixture rapidly, without stirring, for 10 minutes, or until it reaches setting point. Continue to skim off the stones from the top. Remove the pan from the heat to test. The sugar thermometer should read 104°C (220°F). If you do not have a sugar thermometer, test for set with the cold plate test (see page 15).

5 With the pan off the heat, lightly skim off any remaining stones and any scum from the surface of the jam.

6 Immediately pour the jam into warmed sterilized jars, to within 3 mm of the tops. Seal the jars and label.

Makes about 2.25 kg

WARMING SUGAR Sugar will dissolve quickly in the fruit mixture if it is warmed first. Put the oven on its lowest setting. Weigh the sugar and put it into an ovenproof bowl. Warm the sugar in the oven for about 15 minutes. The sterilized jars can be put into the oven to warm at the same time.

Crushed Strawberry Jam

Ingredients

1.9 kg strawberries

90 ml lemon juice

1.8 kg sugar, warmed (see box, page 20)

1 Hull the strawberries and check their weight. You should have 1.8 kg hulled fruit. Halve the strawberries and put into a non-metallic bowl. Lightly crush them with a potato masher.

2 Put the crushed strawberries and lemon juice into a preserving pan and bring to a boil. Lower the heat and simmer for 5–10 minutes, or until the strawberries are soft.

3 Add the warmed sugar to the strawberry mixture, and stir over a low heat, until the sugar has completely dissolved.

4 Increase the heat and boil the mixture rapidly, without stirring, for 15 minutes, or until it reaches setting point. Remove the pan from the heat to test. The sugar thermometer should read 104°C (220°F). If you do not have a sugar thermometer, test for set with the cold plate test (see page 15).

5 With the pan off the heat, lightly skim off any scum from the surface of the jam, using a long-handled metal spoon. Cool the jam slightly.

6 Pour the jam into warmed sterilized jars, to within 3 mm of the tops. Seal the jars and label.

Makes about 2.25 kg

Whole Strawberry Jam Leave whole hulled strawberries to stand with the lemon juice and cold sugar overnight. Put the mixture into a preserving pan, heat gently, and continue as directed from step 4. Leave for 15 minutes before putting into jars.

Brandied Carrot Jam

Ingredients

1 kg young carrots

900 ml water

675 g sugar, warmed (see box, page 20)

finely grated zest and juice of 2 large lemons

3 tsp freshly grated root ginger

1½ tbsp brandy

1 Top and tail, scrape, and coarsely chop the carrots. Put the carrots and water into a preserving pan and bring to a boil. Lower the heat, and simmer, covered, for 20 minutes, or until the carrots are very soft. Purée the carrots in a food processor or blender. Alternatively, press them through a plastic sieve.

2 Return the carrots to the pan. Add the warmed sugar, lemon zest and juice, and grated ginger to the carrot mixture. Stir over a low heat, until the sugar has completely dissolved.

3 Increase the heat and boil the mixture rapidly, without stirring, for 15–20 minutes, or until it reaches setting point. Remove the pan from the heat to test. The sugar thermometer should read 104°C (220°F). If you do not have a sugar thermometer, test for set with the cold plate test (see page 15).

4 With the pan off the heat, skim off any scum from the surface of the jam, using a long-handled metal spoon. Stir in the brandy.

5 Immediately pour the jam into warmed sterilized jars, to within 3 mm of the tops. Seal the jars and label.

Makes about 575 g

BLUEBERRY JAM

INGREDIENTS

1.8 kg blueberries

1.4 kg sugar

juice of 2 lemons

pinch of salt

Sweet spicy blueberries are a North American favourite, and are distant relatives of the European bilberry.

1 Put the blueberries into a non-metallic bowl with half of the sugar, all of the lemon juice, and the salt. Stir to mix, cover, and leave to stand for about 5 hours.

2 Pour the contents of the bowl into a preserving pan. Add the remaining sugar, and stir over a low heat, with a wooden spoon, until the sugar has completely dissolved.

3 Increase the heat and boil the mixture rapidly, without stirring, for 10–12 minutes, or until it reaches setting point. Remove the pan from the heat to test. The sugar thermometer should read 104°C (220°F). Alternatively, test for set with the cold plate test (see page 15).

4 With the pan off the heat, skim off any scum from the surface of the jam, using a long-handled metal spoon.

5 Immediately pour the jam into warmed sterilized jars, to within 3 mm of the tops. Seal the jars and label.

Makes about 3.25 kg

RASPBERRY AND REDCURRANT JAM

INGREDIENTS

225 g redcurrants

1.15 litres water

1.8 kg raspberries

2.25 kg sugar, warmed (see box, page 20)

1 Put the redcurrants into a preserving pan set over a low heat. When the juices start to run, add the water, and bring to a boil. Simmer, stirring, for 15 minutes.

2 Strain the redcurrants through a plastic sieve into a large measuring jug, pressing with the back of a wooden spoon to extract all the juice. Discard the pulp in the sieve. The juice should measure 900 ml so reduce it if necessary by boiling it for longer.

3 Pour the redcurrant juice back into the pan. Add the raspberries and return to a boil. Lower the heat and simmer for 10 minutes. Add the warmed sugar to the mixture, and stir, until the sugar has completely dissolved.

4 Increase the heat and boil the mixture rapidly, without stirring, for 10–15 minutes, or until it reaches setting point. Remove the pan from the heat to test. The sugar thermometer should read 104°C (220°F). Alternatively, test for set with the cold plate test (see page 15).

5 With the pan off the heat, lightly skim off any scum from the surface of the jam, using a long-handled metal spoon. Immediately pour the jam into warmed sterilized jars, to within 3 mm of the tops. Seal the jars and label.

Makes about 3.25 kg

From left to right, *Blueberry Jam, Pineapple Jam, and Raspberry and Redcurrant Jam*

Blackberry and Apple Jam

INGREDIENTS

1 kg cooking apples

300 ml water

1.8 kg blackberries

2.75 kg sugar, warmed (see box, page 20)

1 Peel, core, and roughly chop the apples. You should have about 675 g in weight. Put the apples and half of the water into a medium saucepan. Bring to a boil, lower the heat, and simmer, stirring occasionally, for about 10 minutes, or until the apples are quite soft. Set aside.

2 Put the blackberries and the remaining water into a preserving pan. Bring to a boil, and simmer for about 15 minutes, or until the blackberries are soft. Add the apples and return to a boil. Add the warmed sugar to the fruit mixture, and stir with a wooden spoon until the sugar has completely dissolved.

3 Increase the heat and boil the mixture rapidly, without stirring, for about 10 minutes, or until it reaches setting point. Remove the pan from the heat to test. The sugar thermometer should read 104°C (220°F). If you do not have a sugar thermometer test for set with the cold plate test (see page 15).

4 With the pan off the heat, lightly skim off any scum from the surface of the jam, using a long-handled metal spoon. Immediately pour the jam into warmed sterilized jars, to within 3 mm of the tops. Seal the jars and label.

Makes about 4.5 kg

Pineapple Jam

INGREDIENTS

2 ripe pineapples, total weight about 1.4 kg with leaves

sugar

about 2 lemons

1 Slice the tops off the pineapples. Cut off the peel in strips, cutting deep enough to remove the "eyes" with the peel. Cut the pineapples across into thick slices, cut out and discard the hard central core from each slice; dice the flesh. Weigh the fruit. For each 450 g fruit, weigh 450 g sugar. Put the fruit into a saucepan. Warm the sugar (see box, page 20), and add it to the fruit.

2 Allow 1 lemon for each 450 g fruit. Peel off the zest and slice it finely. Halve and squeeze the lemons, keeping the pips and juice. Roughly chop the squeezed lemon halves. Put them with the pips on a square of muslin and tie up tightly into a bag with a long piece of string.

3 Put the lemon zest and juice into the pan with the pineapple and tie the muslin bag to the pan handle. Stir over a low heat, until the sugar has completely dissolved Increase the heat and bring to a boil, then simmer, stirring occasionally, for about 1½ hours. Remove the pan from the heat. Test for set with the cold plate test (see page 15). It will be a lighter set than most jams. Discard the bag, squeezing it first to extract all the juice.

4 With the pan off the heat, lightly skim off any scum from the surface of the jam, using a long-handled metal spoon. Immediately pour the jam into warmed sterilized jars, to within 3 mm of the tops. Seal the jars and label.

Makes about 1.4 kg

NECTARINE JAM

INGREDIENTS

1 kg nectarines

240 ml water

675 g sugar

75 ml lemon juice

120 g package liquid fruit pectin

Nectarines, like peaches, are low in pectin, so it is easier to achieve a set with this jam if a commercial liquid fruit pectin is added. When using this pectin, lemon juice must be added and the jam must not cook for too long because it reaches setting point quickly and overcooking reduces the pectin setting properties.

1 Halve and stone the nectarines, but do not peel them. Chop them coarsely.

2 Put the chopped nectarines and water into a preserving pan, and simmer for 15 minutes, or until the fruit is soft.

3 Crush the chopped nectarines with a potato masher. Add the sugar, lemon juice and pectin to the nectarine mixture, and stir over a low heat, until the sugar has completely dissolved.

4 Increase the heat and boil the mixture rapidly, without stirring, for only 1 minute or according to package instructions.

5 With the pan off the heat, lightly skim off any scum from the surface of the jam, using a long-handled metal spoon. Cool the jam slightly.

6 Pour the jam into warmed sterilized jars, to within 3 mm of the tops. Seal the jars and label.

Makes about 675 g

COOK'S TIP Use a small sharp knife to remove stones from fruits. Using the indentation on one side as a guide, cut the fruit in half. Then, with both hands, give a sharp twist to each half to loosen the stone. Lift or scoop out the stone with the knife.

LOGANBERRY JAM

INGREDIENTS

2.75 kg loganberries

juice of 1 lemon

2.75 kg sugar, warmed (see box, page 20)

1 Put the loganberries and lemon juice into a preserving pan. Simmer over a medium heat, stirring, until the juices run free. Bring the mixture to a boil, then lower the heat and simmer, stirring occasionally, for 15–20 minutes, or until the fruit is soft.

2 Add the warmed sugar to the loganberry mixture, and stir with a wooden spoon, until the sugar has completely dissolved.

3 Increase the heat and boil the mixture rapidly, without stirring, for 4–6 minutes, or until it reaches setting point. Remove the pan from the heat to test. The sugar thermometer should read 104°C (220°F). If you do not have a sugar thermometer, test for set with the cold plate test (see page 15).

4 With the pan off the heat, lightly skim off any scum from the surface of the jam, using a long-handled metal spoon.

5 Immediately pour the jam into warmed sterilized jars, to within 3 mm of the tops. Seal the jars and label.

Makes about 4.5 kg

VARIATIONS
Raspberries can be used in place of loganberries. Alternatively, try using dewberries, boysenberries, blackberries, even a mixture of berries. Wild or freshly picked berries give jams the best flavour.

RHUBARB AND STRAWBERRY JAM

Rhubarb and strawberries combined together make a wonderfully scented chunky jam.

INGREDIENTS

1.4 kg rhubarb

450 g strawberries

1.4 kg sugar

3 lemons

1 Trim the rhubarb and cut the stalks into 1.25 cm pieces. Hull and halve the strawberries. Layer the fruit pieces with the sugar in a large non-metallic bowl. Halve and squeeze the lemons, keeping the pips, juice, and lemon halves. Pour the juice over the layered fruit. Cover and leave to stand overnight, to draw out the juices.

2 Roughly chop the squeezed lemon halves. Put them with the pips on a square of muslin and tie up tightly into a bag with a long piece of string. Pour the rhubarb and strawberry mixture into a preserving pan and tie the muslin bag to the pan handle, so that it rests on the fruit.

3 Bring the mixture to a boil over a high heat and boil rapidly, without stirring, for 15 minutes, or until it reaches setting point. Remove the pan from the heat to test. The sugar thermometer should read 104°C (220°F). If you do not have a sugar thermometer, test for set with the cold plate test (see page 15). Lift the bag out of the pan and squeeze all the juice back into the pan. Discard the bag.

4 With the pan off the heat, lightly skim off any scum from the surface of the jam, using a long-handled metal spoon. Immediately pour the jam into warmed sterilized jars, to within 3 mm of the tops. Seal the jars and label.

Makes about 2 kg

Above, *Rhubarb and Strawberry Jam;* ***right,*** *Fresh Apricot Jam*

FRESH APRICOT JAM

INGREDIENTS

2.75 kg apricots

600 ml water

juice of 1 lemon

2.75 kg sugar, warmed (see box, page 20)

1 Halve and stone the apricots. Crack a few of the stones with a hammer and take out the kernels. Discard the rest. Put the kernels into a small bowl and pour over boiling water to cover. Leave for 1 minute, then drain and transfer to a bowl of cold water. Drain again, then rub off the skins with your fingers.

2 Put the apricots, kernels, measured water, and lemon juice into a preserving pan and bring to a boil. Lower the heat and simmer, stirring occasionally, for 20–30 minutes, or until the apricot skins are soft and the fruit is tender. The mixture in the pan should have reduced by about one-third. Add the warmed sugar to the apricot mixture and stir until the sugar has completely dissolved.

3 Increase the heat and boil, without stirring, for 10 minutes, or until it reaches setting point. Remove the pan from the heat to test. The sugar thermometer should read 104°C (220°F). Alternatively, test for set with the cold plate test (see page 15).

4 With the pan off the heat, lightly skim off any scum from the surface of the jam, using a long-handled metal spoon. Immediately pour the jam into warmed sterilized jars, to within 3 mm of the tops. Seal the jars and label.

Makes about 4.5 kg

Blackcurrant Jam

INGREDIENTS

1.8 kg blackcurrants

1.4 litres water

2.75 kg sugar, warmed (see box, page 20)

Blackcurrants are the ideal fruit for jam making, because they have a high level of pectin. Remove them from their stalks with a fork.

1 Put the blackcurrants and water into a preserving pan and bring to a boil. Lower the heat and simmer, stirring occasionally with a wooden spoon, for 50–60 minutes, or until the blackcurrant skins are soft and the fruit is tender. The fruit mixture should have reduced by about one-third.

2 Add the warmed sugar to the blackcurrant mixture, and stir over a low heat, until the sugar has completely dissolved.

3 Increase the heat and boil the mixture rapidly, without stirring, for 6–8 minutes, or until it reaches setting point. (This jam reaches setting point quickly, so start testing early.) Remove the pan from the heat to test. The sugar thermometer should read 104°C (220°F). If you do not have a sugar thermometer, test for set with the cold plate test (see page 15).

4 With the pan off the heat, lightly skim off any scum from the surface of the jam, using a long-handled metal spoon.

5 Immediately pour the jam into warmed sterilized jars, to within 3 mm of the tops. Seal the jars and label.

Makes about 3.25 kg

Plum Jam

INGREDIENTS

2 kg sweet green plums

450 ml water

1.8 kg sugar, warmed (see box, page 20)

1 Halve and stone the plums and check their weight. You should have 1.8 kg stoned fruit. Crack a few of the stones with a hammer and take out the kernels. Discard the rest. Put the kernels into a small bowl and pour over boiling water to cover. Leave for 1 minute, then drain and transfer to a bowl of cold water. Drain again, then rub off the skins with your fingers.

2 Put the plums, kernels, and measured water into a copper or brass pan and bring to a boil. Lower the heat and simmer, stirring occasionally, for 30–40 minutes, or until the plum skins are soft and the fruit is tender. The mixture in the pan should have reduced by about one-third.

3 Add the warmed sugar to the plum mixture, and stir over a low heat, until the sugar has completely dissolved.

4 Increase the heat and boil the mixture rapidly, without stirring, for about 10 minutes, or until it reaches setting point. Remove the pan from the heat to test. The sugar thermometer should read 104°C (220°F). If you do not have a sugar thermometer, test for set with the cold plate test (see page 15).

5 With the pan off the heat, lightly skim off any scum from the surface of the jam, using a long-handled metal spoon.

6 Immediately pour the jam into warmed sterilized jars, to within 3 mm of the tops. Seal the jars and label.

Makes about 2.25 kg

DID YOU KNOW? Jams made with green plums and gooseberries tend to discolour when made in a stainless steel pan. Use a copper or brass pan instead.

Apple and Ginger Jam

INGREDIENTS

2.75 kg cooking apples

1.15 litres water

2 tsp ground ginger

finely grated zest and juice of 4 lemons

75 g crystallized ginger or stem ginger in syrup

2.75 kg sugar, warmed (see box, page 20)

1 Peel, core, and roughly chop the apples. Put the peel and cores on a square of muslin and tie up tightly into a bag with a long piece of string. Put the apples and water into a preserving pan with the ground ginger, lemon zest and juice. Tie the muslin bag to the pan handle. Bring to a boil, lower the heat and simmer, stirring occasionally, for 10 minutes, or until the apples are soft. Discard the bag, squeezing it first to extract all the juice.

2 Chop the crystallized or stem ginger; set aside. Add the warmed sugar to the apples, and stir over a low heat, until the sugar has completely dissolved.

3 Increase the heat and boil the mixture rapidly, without stirring, for 10 minutes, or until it reaches setting point. Remove the pan from the heat to test. The sugar thermometer should read 104°C (220°F). If you do not have a sugar thermometer, test for set with the cold plate test (see page 15).

4 With the pan off the heat, lightly skim off any scum from the surface of the jam, using a long-handled metal spoon. Stir in the ginger, and pour the jam into warmed sterilized jars, to within 3 mm of the tops. Seal the jars and label.

Makes about 4.15 kg

Gooseberry Jam

INGREDIENTS

1.8 kg gooseberries

600 ml water

5 large elderflowers (optional)

2.25 kg sugar, warmed (see box, page 20)

1 Top and tail the gooseberries. Put them with the water into a copper or brass preserving pan (see box, page 26).

2 If using elderflowers, cut off and discard the stems and wash the heads. Put them on a large square of muslin and tie up tightly into a bag with a long piece of string. Tie the muslin bag to the pan handle, so that it rests on the fruit, and bring the fruit and flower mixture to a boil. Lower the heat and simmer, stirring occasionally, for 30–40 minutes, or until the fruit is soft. The mixture in the pan should have reduced by about one-third. Discard the bag, if used, squeezing it first to extract all of the juice.

3 Add the warmed sugar to the gooseberry mixture, and stir over a low heat, until the sugar has completely dissolved.

4 Increase the heat and boil the mixture rapidly, without stirring, for 6–8 minutes, or until it reaches setting point. (This jam reaches setting point quickly, so start testing early.) Remove the pan from the heat to test. The sugar thermometer should read 104°C (220°F). If you do not have a sugar thermometer, test for set with the cold plate test (see page 15).

5 With the pan off the heat, lightly skim off any scum from the surface of the jam, using a long-handled metal spoon.

6 Immediately pour the jam into warmed sterilized jars, to within 3 mm of the tops. Seal the jars and label.

Makes about 3.25 kg

Preserves & Conserves

Fruit preserves, in the narrow sense of the word, are whole fruits or pieces of fruit preserved in a thick syrup – a definition that is accurate but cannot possibly convey the deliciousness of biting into fruits captured from summer's bounty by the subtle use of sugar.

Almost any fruits can be preserved for our enjoyment in this very simple way – peaches, pears, and strawberries are suitable, so too are exotics like melons and kiwi fruit. You can preserve each type of fruit on its own, or mix several different ones together for extra flavour and interest. Serve preserves by themselves, as a dessert topping with ice-cream, or on a sponge flan base topped with fresh fruit and whipped cream.

A close culinary cousin of the preserve is the conserve, best described as a type of jam. Conserves are often made with two or more fruits, one of which may be a citrus fruit, and usually raisins or nuts are added. They were eaten on their own as a dessert course in Victorian and Edwardian days, but these days they are more usually served as spreads, like jams and jellies, or as dessert sauces. They also make fine accompaniments to savoury dishes.

Making Preserves & Conserves

❦ Choose the ingredients. All fresh fruits should be picked or purchased when slightly under-ripe. Dried fruits – prunes, raisins, and apricots – can also be used, provided that they have not been preserved with sulphur dioxide, because this affects the set. Any dried fruits should be soaked overnight before using in these recipes.

❦ Prepare the fruits. Pick over soft fruits, removing and discarding any mouldy or damaged parts. Cut out any blemishes from firmer fruits, such as pears. Wash the fruits, using as little cold water as possible for soft fruits. Pat the fruits dry on kitchen paper to remove all excess water. Stone fruits should be halved and stoned, and a few kernels reserved if the recipe calls for it.

❦ For preserves, the chosen fruit is layered with sugar in a non-metallic bowl and left to stand overnight. This helps toughen the fruit, enabling it to stay whole, and it also draws out the juices. The next day, the fruit, sugar, and extracted juices are boiled together briefly until a thick syrup is formed. Cool until the fruit remains suspended in the syrup.

❦ The method of making conserves is very similar to jams but they have a shorter cooking time, so they retain their natural flavour better. The fruit is cooked over a low heat until the sugar completely dissolves, and then boiled fast to reach a light set.

❦ To test for a light set, drop a little mixture on a cold plate and chill quickly; if the conserve forms a slight wrinkle when pushed with a finger, it is ready.

❦ If adding nuts or alcohol, such as blanched or flaked almonds or Grand Marnier, then add them at the last minute, because cooking them will destroy their flavour. Pour the preserves or conserves into warmed sterilized jars (see page 11).

How to seal and store

Seal the preserve or conserve as soon as it has been put into the jar. Choose screw-top lids with plastic linings that can be secured tightly. Label the jars. Preserves and conserves do not keep as long as jam, and should be eaten within 3 months. To keep them longer, store them in the refrigerator. They should then last about 6 months.

What can go wrong and why

If the fruit does not stay whole in a preserve, it has not been left overnight with the sugar, or it has been cooked too long. If the preserve or conserve shrinks in the jar, the seal is faulty or the storage conditions are too warm. Air pockets in the jar are caused if the preserve or conserve is too cool before being poured into the jar. Problems that occur in jams also apply to these recipes (see page 15).

Prune Conserve

INGREDIENTS

675 g prunes

675 g raisins

675 g currants

1.15 litres hot strong tea

450 g brown sugar

3–5 whole cloves

juice of 1 lemon

100 g whole blanched almonds

100 ml brandy or kirsch

1 Remove the stones from the prunes and chop the prunes roughly. Crack a few of the stones with a hammer and take out the kernels. Discard the remaining stones.

2 Put the kernels into a small bowl and pour boiling water over to cover. Leave for 1 minute, then drain and transfer to a bowl of cold water. Drain again, then rub off the skins with your fingers.

3 Put the chopped prunes and peeled kernels into a non-metallic bowl with the raisins and currants. Pour the hot tea over the fruit. Cover and leave to stand overnight.

4 Pour the contents of the bowl into a preserving pan. Stir in the sugar, cloves, and lemon juice and put the pan over a low heat. Stir with a wooden spoon until the sugar has completely dissolved.

5 Bring the mixture to a boil. Simmer, stirring constantly, for 12 minutes, or until the mixture has thickened. Remove the pan from the heat. Test for a light set (see page 28). Stir in the whole almonds and brandy or kirsch.

6 Pour the conserve into warmed sterilized jars, to within 3 mm of the tops. Seal the jars and label.

Makes about 3.25 kg

Peach and Melon Preserve

INGREDIENTS

1.15 kg peaches

1.4 kg jam melons

1.4 kg sugar

2 1/2 tbsp lemon juice

pinch of ground ginger

100 g whole blanched almonds

1 Stone, peel, and roughly chop the peaches. Peel, seed, and chop the melons into large chunks. You should have 1 kg each of chopped peach and melon.

2 Layer the fruits with the sugar in a large non-metallic bowl. Cover and leave to stand overnight.

3 Pour the contents of the bowl into a preserving pan and add the lemon juice and ground ginger. Stir the mixture over a low heat until the sugar has completely dissolved.

4 Increase the heat and boil the mixture rapidly for 20 minutes, or until the mixture has thickened.

5 Remove the pan from the heat and cool slightly until the fruit remains suspended in the syrup. Coarsely chop the almonds and gently stir into the mixture. Pour the preserve into warmed sterilized jars, to within 3 mm of the tops. Seal the jars and label.

Makes about 2.75 kg

Above, *Peach and Melon Preserve;* **above right,** *Prune Conserve*

STRAWBERRY PRESERVE

INGREDIENTS

1.8 kg small strawberries

1.8 kg sugar

juice of 1 lemon

Here, whole strawberries are suspended in a sugar syrup.

1 Hull the strawberries. Layer them with the sugar in a non-metallic bowl. Cover and leave to stand overnight.

2 Pour the contents of the bowl into a preserving pan and add the lemon juice. Bring the mixture to a boil, and simmer for 5 minutes. Return the strawberry mixture to the bowl, cover, and leave for 24 hours.

3 Return the strawberry mixture to the pan and boil rapidly for 20–25 minutes, or until the syrup has thickened. Cool slightly until the strawberries remain suspended in the syrup. Pour the preserve into warmed sterilized jars, to within 3 mm of the tops. Seal the jars and label.

Makes about 2.25 kg

COOK'S TIP Strawberry Preserve can be used to make strawberry pavlova – spread a layer of the preserve carefully in a baked meringue shell, and top with whipped cream and fresh strawberries.

DRIED APRICOT CONSERVE

INGREDIENTS

1 kg dried apricots

2.25 litres water

juice of 2 lemons

1.8 kg sugar, warmed (see box, page 36)

175 g blanched almonds

1 Coarsely chop the apricots. Put them into a non-metallic bowl with the water. Cover and leave to stand overnight.

2 Pour the contents of the bowl into a preserving pan and add the lemon juice. Bring to a boil, and simmer for 15–20 minutes, or until the apricots are really soft. Add the warmed sugar. Stir the mixture over a low heat, with a wooden spoon, until the sugar has completely dissolved.

3 Boil the mixture rapidly, without stirring, for 10–12 minutes, or until it has thickened. Remove the pan from the heat to test. Test for a light set (see page 28). Stir in the blanched almonds and cool slightly.

4 Pour the conserve into warmed sterilized jars, to within 3 mm of the tops. Seal the jars and label.

Makes about 3.25 kg

GOOSEBERRY AND ALMOND CONSERVE

INGREDIENTS

1 kg gooseberries

175 ml lemon juice

300 ml water

1 kg sugar, warmed (see box, page 36)

100 g flaked almonds

1 Top and tail the gooseberries.

2 Combine the gooseberries, lemon juice, and water in a copper or brass pan (see box, page 26). Add the warmed sugar, and stir over a low heat, until the sugar has completely dissolved.

3 Boil the mixture rapidly, stirring occasionally, for 15–20 minutes, or until it has thickened. Remove the pan from the heat to test. Test for a light set (see page 28). Stir in the flaked almonds and cool slightly.

4 Pour the conserve into warmed sterilized jars, to within 3 mm of the tops. Seal the jars and label.

Makes about 1.4 kg

MELON AND ORANGE PRESERVE

INGREDIENTS

2.1 kg jam melons

775 g sugar

finely grated zest and juice of 1 large orange

juice of 1 lemon

1 Peel and seed the melons, cut into 1.25 cm cubes, and check the weight. You should have 1.4 kg melon flesh. Layer the melon with the sugar in a non-metallic bowl. Cover and leave to stand overnight.

2 Pour the contents of the bowl into a preserving pan and add the grated orange zest and the orange and lemon juice.

3 Bring the mixture to a boil over a high heat, and boil rapidly for 20–25 minutes, or until the syrup has reduced. Cool slightly until the melon remains suspended in the syrup.

4 Pour the preserve into warmed sterilized jars, to within 3 mm of the tops. Seal the jars and label.

Makes about 1.4 kg

PEAR AND CASHEW CONSERVE

INGREDIENTS

1 kg pears

300 ml water

juice of 1 lemon

1.15 kg sugar, warmed (see box, page 36)

75 g cashew nuts

1 Peel, halve, and core the pears, and cut the flesh into cubes. Put the pears into a saucepan with the water and lemon juice. Bring to boil and simmer for 10 minutes, or until the pears are soft and tender.

2 Add the warmed sugar to the fruit mixture and stir over a low heat until the sugar has completely dissolved. Boil rapidly for 15–20 minutes, or until the mixture has thickened. Remove the pan from the heat. Test for a light set (see page 28). Chop the cashews and stir them into the conserve. Cool slightly.

3 Pour the conserve into warmed sterilized jars, to within 3 mm of the tops. Seal the jars and label.

Makes about 1.4 kg

VARIATION

Omit the cashew nuts. Add 3 tsp freshly grated root ginger to the pears in the pan in step 1. When the pear preserve is cool, stir in 3 tsp rum.

PLUM AND ORANGE CONSERVE

INGREDIENTS

1.4 kg yellow plums

100 g stem ginger in syrup, drained

100 g walnuts

300 ml water

finely grated zest and juice of 2 large oranges and 2 lemons

1.8 kg sugar, warmed (see box, page 36)

450 g raisins

2 tbsp brandy

1 Halve and stone the plums. Coarsely chop the drained stem ginger and the walnuts; set them aside separately.

2 Put the plum halves into a preserving pan with the water and the zest and juice of the oranges and lemons. Bring to a boil, and simmer for 30 minutes.

3 Add the warmed sugar to the fruit mixture, and stir over a low heat until the sugar has completely dissolved. Add the chopped ginger and the raisins, and return to a boil.

4 Boil rapidly until the mixture has thickened. Remove the pan from the heat. Test for a light set (see page 28). Stir in the walnuts and brandy. Cool slightly.

5 Pour the conserve into warmed sterilized jars, to within 3 mm of the tops. Seal the jars and label.

Makes about 3.5 kg

MARMALADES

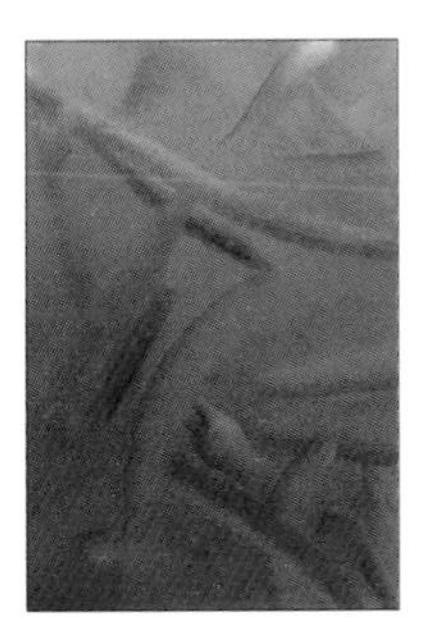

IN VICTORIAN and Edwardian England, marmalade was a fruit purée that was eaten with a spoon out of an elegant small dish – and it still is served in this way in some parts of the world, including Latin America and France. But generally, marmalade is the delectable, golden wonder we make from citrus fruits and serve on toast at breakfast time.

This preserve is beautiful to look at in jars, and just about any citrus fruit can be used to make it. Marmalade can be dark coloured, thick, and chunky, or a translucent lemon jelly with thin slices of fruit. In fact, with so many possible combinations of fruit, it can be anything in between.

A citrus preserve is not just for spreading on toast. It can be mixed into puddings and cakes, or melted and used with a little liqueur to make a marmalade sauce. In Scotland, marmalade is spread on scones and oatcakes, as well as being eaten with certain meats. Make lots and keep storecupboards well stocked, because marmalade keeps exceedingly well.

MAKING MARMALADES

❦ Select the ingredients. Bitter Seville oranges give the tangiest, clearest marmalade, but their season of availability is quite short, only in winter. They can be frozen until required, but are best used within a few months, because freezing can cause their pectin level to drop slightly. Sweet oranges can also be used, but they tend to make a cloudier preserve, and are best used in combination with other fruits, such as limes, lemons, or grapefruit. White sugar is the recommended sweetener. Sometimes the exact amount of sugar is not calculated until after the fruits have been cooked, so make sure you have plenty in stock before you start.

❦ Prepare the fruits. Some citrus fruits have a wax coating which should be removed by pouring boiling water over the fruits, scrubbing, and drying them. The way the fruits are prepared depends on the type of marmalade you are making. There are two basic types – jelly-like, or thick and chunky. Either slice the peel or zest by hand, or speed the job up by using a food processor, though this does not give such an even cut. Most of the pectin in citrus fruits, which is needed to set the marmalade, is found in the pith and pips. They should be tied up in a muslin bag and cooked with the rest of the ingredients.

❦ Put the sliced peel or zest, juice, specified amount of water, lemon juice if required, and muslin bag into a preserving pan and simmer for 1 hour or more, until the liquid is reduced by about one-third. This is to soften the peel thoroughly, and to extract all of the pectin. Remove the muslin bag, squeezing it out well to extract all the excess liquid.

❦ Add warmed sugar (see box, page 36), and stir until dissolved. Boil rapidly until setting point is reached. Test for set and skim (see box, page 38). Allow the marmalade to stand for a few minutes off the heat. This helps to ensure that the peel is evenly distributed throughout. As soon as a thin skin forms on the surface, stir the marmalade once, and then pour into warmed sterilized jars (see page 11).

How to seal and store

Seal the marmalade as for jams (see page 15), putting the lids on when the marmalade is cold. Label the jars. The marmalade can be used immediately, or it will keep in a cool dark place for up to 1 year.

What can go wrong and why

If the peel is tough, it has been shredded too coarsely, or cooked for too short a time before the sugar is added. If the peel rises to the surface, the jars were too hot when the marmalade was poured into them.

SEVILLE ORANGE MARMALADE

INGREDIENTS

1 kg Seville oranges

1 large lemon

2.25 litres water

1.8 kg sugar, warmed
(see box, page 36)

1 Cut all the fruit in half and squeeze out the juice, removing and keeping the pips and any excess pith. Strain the juice into a preserving pan. Cut the orange halves in half again and slice thinly across their length. Put the sliced orange peel into the preserving pan.

2 Roughly chop the lemon halves. Put the pips, pith, and chopped lemon halves on a square of muslin and tie up tightly into a bag with a long piece of string. Tie the muslin bag to the pan handle, so that it rests on the fruit.

3 Pour in the water and bring to a boil. Simmer, stirring occasionally, for at least 2 hours, or until the peel is very soft. The mixture should have reduced by about one-third.

4 Lift the bag out of the pan and squeeze all the juice back into the pan; discard the bag. Add the warmed sugar and stir over a low heat until it has completely dissolved.

5 Increase the heat and boil rapidly, without stirring, for 15–20 minutes, or until the marmalade reaches setting point. Test for set and skim off any scum (see box, page 38).

6 Pour the marmalade into warmed sterilized jars, to within 3 mm of the tops. Seal and label.

Makes about 3.5 kg

SEVILLE ORANGES The bitter flavour of Seville oranges makes them one of the best citrus fruits for making marmalade. Unfortunately, they are only available for a limited period, but this recipe, or any other specifying Seville oranges, can be made by substituting sweet oranges for Seville. Just follow this simple formula – for each 450 g Seville oranges called for, weigh the same amount of sweet oranges. Remove one of the oranges and replace with 1 lemon. The lemon should be dealt with in the same way as the oranges.

MORE MARMALADE FLAVOURS

The popularity of Seville Orange Marmalade makes it a good basic recipe that can be varied by adding other flavourings. Here are a few suggestions.

Liqueur Marmalade
A few tablespoons of rum, brandy, or whisky added to marmalade turn it into a luxurious breakfast preserve. Simply follow the basic recipe above, then stir in the spirit after skimming the marmalade. Allow 1½ tbsp spirit for each 450 g marmalade. Let the marmalade stand for a few minutes to ensure the orange peel is evenly distributed throughout. Stir the marmalade once, and continue as directed.

Dark Chunky Marmalade
Follow the basic recipe above, coarsely chopping the orange peel instead of thinly slicing it. When adding the sugar, also stir in 1½ tbsp black treacle or molasses.

Seville Orange Marmalade with Almonds
Add 100 g flaked almonds to the basic recipe, after skimming the marmalade. Allow the marmalade to stand for a few minutes to ensure the orange peel and nuts are evenly distributed throughout. Stir the marmalade once, and continue as directed.

Brown Sugar Marmalade
For a dark colour and rich flavour, substitute 1.4 kg brown sugar for the white sugar in the basic recipe above.

DID YOU KNOW?
Marmalade has a long history. Its origin stems from the Middle Ages, when it was made with quinces cooked in honey, wine, and spices, and called "marmelade". The word is derived from *marmelada,* Portuguese for quince. It was during the seventeenth century that recipes for a kind of orange marmalade appeared.

THREE-FRUIT MARMALADE

Wonderfully tangy, this marmalade can be made all year round because it does not need seasonal Seville oranges. Don't save it just for breakfast – try using it as a glossy glaze for brushing over duck, bacon rashers, and chicken breasts.

INGREDIENTS

2 grapefruit

2 large oranges

4 lemons

3.5 litres water

2.75 kg sugar, warmed (see box, page 36)

Makes about 4.75 kg

1 ◄ Cut all of the fruit in half and squeeze out the juice. Strain the juice into a preserving pan. Remove the pips and pulp from the skins and put them on a square of muslin. Tie up tightly into a bag with a long piece of string.

2 ► Cut the orange and lemon halves in half again and slice thinly across their length. Cut each grapefruit half into 4 pieces and slice in the same way. Put all the sliced fruit peel into the pan.

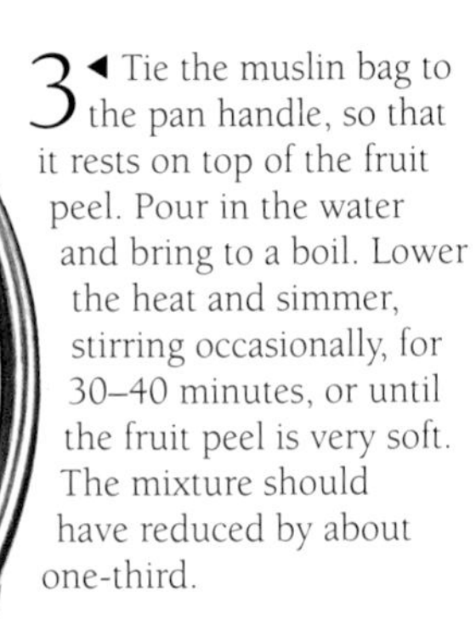

3 ◄ Tie the muslin bag to the pan handle, so that it rests on top of the fruit peel. Pour in the water and bring to a boil. Lower the heat and simmer, stirring occasionally, for 30–40 minutes, or until the fruit peel is very soft. The mixture should have reduced by about one-third.

4 ◀ Lift the bag out of the pan and squeeze all the juice back into pan. Discard the bag. Add the warmed sugar and stir over a low heat, with a wooden spoon, until the sugar has completely dissolved.

5 ▲ Increase the heat and boil the mixture rapidly, without stirring, for 20–25 minutes, or until it reaches setting point. Remove from the heat to test. The sugar thermometer should read 104°C (220°F). Alternatively, test for set with the cold plate test (see box, below).

COLD PLATE TEST
Also known as the saucer test. Put a small spoonful of marmalade on a cold plate and chill quickly in the refrigerator. If the marmalade has boiled sufficiently, a thin skin will form on the surface and it should wrinkle when pushed with a finger. It is now at setting point.

6 ▲ With the pan off the heat, skim off any scum. Leave to cool for a few minutes until a very thin skin forms. This helps to ensure that the peel is evenly distributed.

7 Pour the marmalade into warmed sterilized jars, to within 3 mm of the tops. Seal the jars and label.

BREAKFAST TREAT
Start the day with the refreshing taste of homemade marmalade.

FINE SHRED MARMALADE

INGREDIENTS

1 kg Seville oranges

1 lemon

2.5 litres water

sugar

1 Pare the zest from the oranges and lemon and slice it into thin strips. Put the sliced zest into a stainless steel pan with 300 ml of the water, making sure the zest is covered, and bring to a boil. Simmer, stirring occasionally, for 1 hour, or until the zest is very soft. Drain and reserve both zest and the water.

2 Chop the fruit and put into a preserving pan with all the pips and the remaining water. Bring to a boil and simmer, stirring occasionally, for about 1 hour. The mixture should have reduced by about one-third. Add the water reserved from cooking the zest.

3 Strain the pulp through a scalded jelly bag (see box, page 41). (For a clear marmalade, allow the juice to drip through without pressing it.) Measure the strained juice. For each 600 ml juice, weigh 450 g sugar. Warm the sugar (see box, left). Pour the juice into the pan, add the sugar, and stir over a low heat until the sugar has completely dissolved.

4 Add the reserved zest and bring to a boil. Boil rapidly, without stirring, for 10–12 minutes, or until the marmalade reaches setting point. Test for set and skim off any scum (see box, page 38). Pour the marmalade into warmed sterilized jars, to within 3 mm of the tops. Seal the jars and label.

Makes about 2.75 kg

WARMING SUGAR In some recipes, the amount of sugar required is calculated according to the amount of strained juice you have after the fruit is cooked. It is therefore difficult to be specific with an exact amount of sugar in the ingredients list, so make sure you have a good supply before you start. After weighing out the sugar, it can be warmed so that it dissolves more quickly in the juice. Put the oven on its lowest setting, and put the sugar in an ovenproof bowl. Warm the sugar in the oven for about 15 minutes.

MANDARIN MARMALADE

INGREDIENTS

1 kg mandarins

1 grapefruit

2 lemons

3 litres water

1.6 kg sugar, warmed (see box, above)

1 Peel the mandarins and slice the peel into thin strips. Put the strips of peel on a square of muslin and tie up tightly into a bag with a long piece of string. Chop the mandarin flesh, removing and keeping the pips. Put the flesh into a preserving pan.

2 Pare the zest from the grapefruit and lemons and slice it into thin strips, removing and keeping any excess pith. Chop the grapefruit and lemon flesh, removing and keeping the pips. Put the zest and chopped flesh into the pan. Put all the pips and pith on a square of muslin and tie up tightly into a bag with a long piece of string. Tie both muslin bags to the pan handle, so that they rest on the fruit.

3 Pour in the water and bring to a boil. Simmer, stirring occasionally, for 45 minutes, or until the zest is very soft. The mixture should have reduced by about one-third. Lift the bag of pips and pith out of the pan and squeeze all the juice back into the pan; discard the bag. Lift out the bag of mandarin peel and empty it into the pan. Add the warmed sugar and stir over a low heat until it has completely dissolved.

4 Increase the heat and boil rapidly, without stirring, for 15–20 minutes, or until the marmalade reaches setting point. Test for set and skim off any scum (see box, page 38). Pour the marmalade into warmed sterilized jars, to within 3 mm of the tops. Seal the jars and label.

Makes about 2.75 kg

PINK GRAPEFRUIT MARMALADE

INGREDIENTS

2 lemons

2 pink grapefruit

2.25 litres water

1.8 kg sugar, warmed (see box, page 36)

1 Pare the zest from the lemons and slice it into thin strips. Put the sliced zest into a preserving pan. Cut the lemons and grapefruit in half and squeeze out the juice, removing and keeping the pips. Add the juice to the lemon zest in the pan.

2 Roughly chop the lemon halves and put with the pips on a square of muslin. Tie up tightly into a bag with a long piece of string. Cut the grapefruit halves into 4 pieces and slice thinly across their length. Add to the pan. Tie the muslin bag to the pan handle, so that it rests on the fruit.

3 Pour in the water and bring to a boil. Lower the heat and simmer, stirring occasionally, for 45 minutes, or until the skins are soft. The mixture should have reduced by about one-third.

4 Lift the bag out of the pan and squeeze all the juice back into the pan. Discard the bag. Add the warmed sugar and stir over a low heat until it has completely dissolved. Increase the heat and boil rapidly, without stirring, for 10–12 minutes, or until the marmalade reaches setting point. Test for set and skim off any scum (see box, page 38). Pour the marmalade into warmed sterilized jars, to within 3 mm of the tops. Seal the jars and label.

Makes about 2.25 kg

LEMON AND LIME MARMALADE

INGREDIENTS

4 medium lemons

4 limes

2.25 litres water

1.8 kg sugar, warmed (see box, page 36)

1 Pare the zest from the lemons and slice it into thin strips. Cut the pith off the lemons, roughly chop it, and set aside. Roughly chop the flesh, keeping the pips and any juice.

2 Cut the limes in half and squeeze out the juice, removing and keeping the pips and excess pith. Cut the lime halves in half again and slice thinly across their length.

3 Put all the pips and pith on a square of muslin and tie up tightly into a bag with a long piece of string.

4 Put all the fruit flesh, juice, sliced lime peel, and lemon zest into a preserving pan. Tie the muslin bag to the pan handle, so that it rests on the fruit. Pour in the water and bring to a boil. Simmer, stirring occasionally, for 1½ hours, or until the skins are very soft. The mixture should have reduced by about one-third.

5 Lift the bag out of the pan and squeeze all the juice back into the pan. Discard the bag. Add the warmed sugar and stir over a low heat until it has completely dissolved. Increase the heat and boil rapidly, without stirring, for 8–10 minutes, or until the marmalade reaches setting point. Test for set and skim off any scum (see box, page 38). Pour the marmalade into warmed sterilized jars, to within 3 mm of the tops. Seal the jars and label.

Makes about 2.75 kg

Front, *Pink Grapefruit Marmalade;* **back,** *Lemon and Lime Marmalade*

SWEET ORANGE MARMALADE

INGREDIENTS

1 kg sweet oranges

2 Seville oranges

1 lemon

2.25 litres water

1.8 kg sugar, warmed (see box, page 36)

This marmalade blends sweet oranges with the more tart Seville oranges, to give a delicious and refreshing taste.

1 Cut all the fruit in half and squeeze out the juice, removing and keeping the pips and any excess pith. Pour the juice into a preserving pan. Put the pips and pith on a square of muslin and tie up tightly into a bag with a long piece of string.

2 Cut the orange halves in half again and slice them thinly across their length. Discard the lemon peel.

3 Put all the sliced peel into the pan. Tie the muslin bag to the pan handle, so that it rests on the fruit. Pour in the water and bring to a boil. Simmer, stirring occasionally, for 1½–2 hours, or until the skins are very soft. The mixture should have reduced by about one-third.

4 Lift the bag out of the pan and squeeze all the juice back into the pan. Discard the bag.

5 Add the warmed sugar and stir over a low heat until it has completely dissolved.

6 Increase the heat and boil rapidly, without stirring, for 15–20 minutes, or until the marmalade reaches setting point. Test for set and skim off any scum (see box, left). Pour the marmalade into warmed sterilized jars, to within 3 mm of the tops. Seal the jars and label.

Makes about 3.25 kg

TESTING FOR SET AND SKIMMING MARMALADES
Remove the pan from the heat. If using a sugar thermometer, it should read 104° C (220° F). If you do not have a thermometer, use the cold plate test: drop a little of the marmalade on a cold plate and chill quickly in the refrigerator. If it forms a skin and wrinkles when pushed with a finger, it should be ready. Lightly skim off any scum from the pan of marmalade, then leave to cool for a few minutes until a very thin skin forms. This helps to ensure that the peel is evenly distributed through the marmalade.

QUICK CHUNKY MARMALADE

INGREDIENTS

6 Seville oranges

2 large sweet oranges

2 large lemons

3.5 litres water

2.75 kg sugar, warmed (see box, page 36)

1 Cut all the fruit in half and squeeze out the juice, removing and keeping the pips and any excess pith. Pour the juice into a preserving pan. Put the pips and pith on a square of muslin and tie up tightly into a bag with a long piece of string.

2 Cut the orange and lemon halves into smaller pieces and chop well in small batches, in a food processor, adding a little of the water, if necessary.

3 Put all the processed fruit into the pan. Tie the muslin bag to the pan handle, so that it rests on the fruit. Pour in the water and bring to a boil. Simmer, stirring occasionally, for 1½–2 hours, or until the skins are very soft. The mixture should have reduced by about one-third.

4 Lift the bag out of the pan and squeeze all the juice back into the pan. Discard the bag. Add the warmed sugar to the mixture and stir over a low heat until it has completely dissolved. Increase the heat and boil rapidly, without stirring, for 15 minutes, or until it reaches setting point. Test for set and skim off any scum (see box, above). Pour the marmalade into warmed sterilized jars, to within 3 mm of the tops. Seal the jars and label.

Makes about 3.8 kg

COOK'S TIP
Chopping citrus fruits in a food processor is much quicker than by hand – although it does give a coarser cut. A blender can also be used for chopping, but the result is more like a paste.

LEMON MARMALADE

INGREDIENTS

9 large lemons

3.5 litres water

sugar

1 Pare the zest from the lemons, and slice it into thin strips. Roughly chop the flesh, keeping the pips and any juice. Put the pips on a square of muslin and tie up tightly into a bag with a long piece of string.

2 Put the chopped flesh, and any juice, with the sliced zest in a preserving pan. Tie the muslin bag to the pan handle, so that it rests on the fruit. Pour in the water and bring to a boil. Lower the heat and simmer, stirring occasionally, for 1–$1\frac{1}{2}$ hours, or until the zest is very soft. The mixture should have reduced by about one-third.

3 Lift the bag out of the pan and squeeze all the juice back into the pan. Discard the bag. Measure the juice. For each 600 ml juice, weigh 450 g sugar. Warm the sugar (see box, page 36). Pour the juice back into the pan, add the sugar, and stir over a low heat until it has completely dissolved. Boil rapidly, without stirring, for 15 minutes, or until it reaches setting point. Test for set and skim off any scum (see box, page 38). Pour the marmalade into warmed sterilized jars, to within 3 mm of the tops. Seal the jars and label.

Makes about 2.75 kg

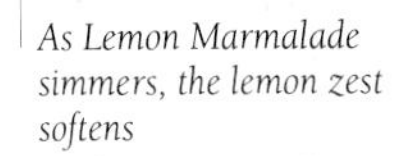

As Lemon Marmalade simmers, the lemon zest softens

GRAPEFRUIT AND GINGER MARMALADE

INGREDIENTS

3 large grapefruit

3 lemons

3.5 litres water

2.5 cm piece of fresh root ginger

2.25 kg sugar, warmed (see box, page 36)

175 g chopped crystallized ginger

1 Peel the grapefruit and lemons. Finely slice the grapefruit peel and put it into a preserving pan. Chop the flesh and add to the pan. Chop the lemon peel and put it on a square of muslin, together with the pips. Tie up tightly into a bag with a long piece of string. Tie the bag to the pan handle, so that it rests on the fruit. Add the water. Peel the root ginger, and add to the pan.

2 Bring to a boil and simmer, stirring occasionally, for $1\frac{1}{2}$ hours, or until the skins are very soft. The mixture should have reduced by about one-third.

3 Remove and discard the ginger. Lift the bag out of the pan and squeeze all the juice back into the pan. Discard the bag. Add the warmed sugar to the mixture and stir over a low heat until the sugar has completely dissolved.

4 Increase the heat and boil rapidly, without stirring, for 15 minutes, or until the marmalade reaches setting point. Test for set and skim off any scum (see box, page 38). Stir in the crystallized ginger. Pour the marmalade into warmed sterilized jars, to within 3 mm of the tops. Seal the jars and label.

Makes about 3.5 kg

Jellies

Brightly coloured, delicately flavoured, firm yet quivering gently when released from its jar, the ideal jelly is a thing of beauty and a joy to eat. It is wonderful on toast, bread and butter, and scones, and in peanut butter sandwiches. It also makes a deliciously sweet counterpoint to savoury foods, such as poultry, pork, and game. And unlike poor Meg's jelly in Louisa May Alcott's book, *Little Women*, today's jellies do jell if the simple rules for making them are respected.

Jellies are a marvellous challenge to the imaginative cook. Peaches, cherries, and cranberries are just a few of the suitable fruits. Apples, with their high levels of pectin which help achieve set, make wonderful companions for other fruits. Apple and Orange Jelly and Apple and Chilli Jelly both testify to the versatility of the apple when making jellies. The addition of a few rose or geranium leaves can lift jellies out of the ordinary, and herbs and spices can also play an interesting role. The possibilities are infinite.

Making Jellies

❦ Choose your ingredients. Select ripe fruits, but not over-ripe, with a high pectin content. Apples (including crab apples), currants, berries, and citrus fruits all make good jellies. Granulated sugar is the recommended sweetener. Make sure you have plenty of sugar in stock, because the exact amount required is not calculated until after the fruits have been cooked.

❦ Prepare the fruits. Pick over soft fruits, removing and discarding any mouldy or damaged parts. Wash and dry the fruits. Any peel, pips, and cores all go in the preserving pan with the flesh. Add water, and simmer until the fruits are soft and pulpy.

❦ Strain the pulp through a scalded jelly bag (see box, page 41). Timing varies, depending on the fruits used. Leave to strain for a few hours or overnight, but for not more than 24 hours or the fruit will start to oxidize.

❦ Measure the juice and pour into a clean preserving pan with the sugar. The amount of sugar used will depend upon the pectin content of the juice, but in general, 450 g sugar is required for each 600 ml juice. Warm the sugar before adding it to the preserving pan (see box, page 44). Stir over a low heat until the sugar has completely dissolved. Boil rapidly for 10 minutes, or until setting point is reached. As the jelly comes up to setting point, lower the heat slightly to reduce the number of bubbles. Remove the pan from the heat to test. Test for set and skim off any scum (see box, page 46).

❦ Skim any scum off the surface, then pour the jelly into warmed sterilized jars (see page 11). Tilt the jar and pour the jelly down the side of the jar to eliminate air pockets. Work as quickly as possible, because the jelly may start to set in the pan and this will spoil the consistency of the finished jelly.

How to seal and store

Seal and store the jellies in the same way as for jams (see page 15). Do not move the jars until the jelly is completely cold and set.

What can go wrong and why

If the jelly remains runny after it is cold, it can be returned to the pan for further boiling. Jellies will be cloudy if the fruits are not clean, if the jelly bag has too coarse a mesh, or if the bag is squeezed during straining. Air pockets will form in the jelly if it is poured into the jars too slowly or if it has been allowed to cool before being poured into the jars.

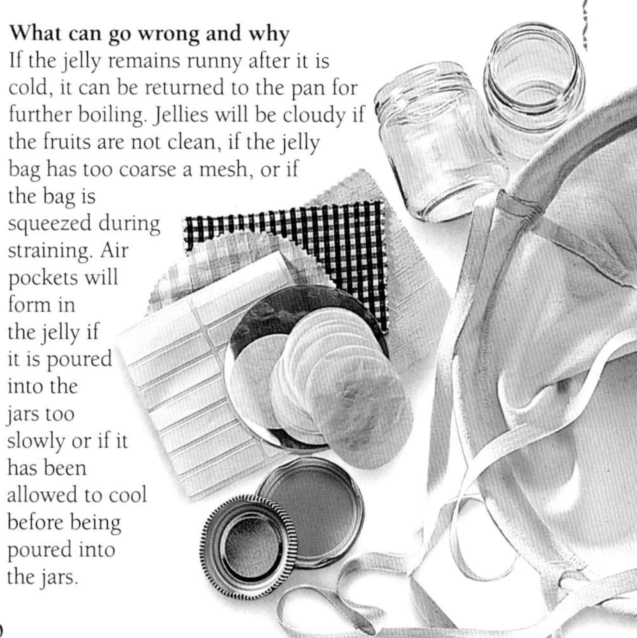

RED FRUIT JELLY WITH PORT

INGREDIENTS

1 kg redcurrants

1 kg raspberries

450 ml water

sugar

3 tbsp port

1 Put the redcurrants and raspberries into a preserving pan with the water. Bring to a boil, and simmer for 15–20 minutes, or until the fruits are very soft and pulpy.

2 While the fruits are cooking, prepare the jelly bag for straining the fruits. Scald it by evenly pouring boiling water through, squeeze it well, then use as directed (see box, left). Pour the fruit pulp into the jelly bag and leave to strain for a few hours or overnight, but for not more than 24 hours.

3 Measure the strained juice. For each 600 ml juice, weigh 450 g sugar. Pour the juice into a clean preserving pan. Warm the sugar (see box, page 44), and add it to the juice. Stir over a low heat until the sugar has completely dissolved.

4 Increase the heat and boil the mixture rapidly, without stirring, for 9–10 minutes, or until it reaches setting point. Test for set and skim off any scum (see box, page 46). Stir in the port.

5 Immediately pour the jelly into warmed sterilized jars, tilting them slightly to prevent air pockets from forming. Seal the jars and label.

Makes about 1.4 kg

USING A JELLY BAG Jelly bags need to be supported while the juice is straining. One way of doing this is to thread a broom handle through the straps of the jelly bag, then rest each end of the broom on a chair. Place a large non-metallic bowl underneath before filling the jelly bag and leave the juice to drip undisturbed. Resist the temptation to accelerate the process by squeezing the bag, because this will result in a cloudy jelly. Leave to drip until there is only a little liquid dripping from the bag.

MINT AND APPLE JELLY

INGREDIENTS

1.4 kg cooking apples

600 ml water

300 ml cider vinegar

225 g fresh mint

sugar

2–3 drops of green food colouring (optional)

1 Chop the apples without peeling or coring them. Put them into a preserving pan with the water and vinegar. Strip the mint leaves from their stalks and set aside. Add the stalks to the pan. Bring the mixture to a boil, and simmer for about 40 minutes, or until the apples are very soft and pulpy.

2 While the apples are cooking, prepare the jelly bag for straining the fruit. Scald it by evenly pouring boiling water through, squeeze it well, then use as directed (see box, above). Pour the fruit pulp into the jelly bag and leave to strain for a few hours or overnight, but for not more than 24 hours.

3 Measure the strained juice. For each 600 ml juice, weigh 450 g sugar. Pour the juice into a clean preserving pan. Warm the sugar (see box, page 44), and add it to the juice. Stir over a low heat until the sugar has completely dissolved. Tie half of the mint leaves in a muslin bag and add to the jelly mixture. Increase the heat and boil the mixture rapidly, without stirring, for 15 minutes, or until it reaches setting point. Test for set and skim off any scum (see box, page 46).

4 Remove the muslin bag, squeezing any juice back into the pan; discard the bag. Chop the remaining mint leaves. Stir in the freshly chopped mint and green food colouring, if using.

5 Immediately pour the jelly into warmed sterilized jars, tilting them slightly to prevent air pockets from forming. Seal the jars and label.

Makes about 675 g

VARIATIONS
Other herb and apple jellies can be made using the same method as for mint. Parsley, thyme, rosemary, and sage are all equally good.

Jelly-making equipment

CRANBERRY AND APPLE JELLY

Ruby-red cranberries are popular for festive occasions in the United States. In this recipe they are teamed up with apples, creating a tart, garnet-coloured jelly. Cranberry and Apple Jelly is a delicious accompaniment to roast pork, cold meats, and poultry – especially roast turkey.

INGREDIENTS

1.4 kg cooking apples

1 kg frozen cranberries

1.15 litres water

about 1 kg sugar

Makes about 1.6 kg

1 ◀ Cut out any bruised or damaged parts from the apples, then coarsely chop the apples without peeling or coring them.

2 ▼ Put the apples into a preserving pan with the frozen cranberries.

3 ▲ Pour in the water to just cover the fruits, and bring to a boil. Lower the heat and simmer for 30 minutes, or until the fruits are very soft and pulpy.

4 While the apples and cranberries are cooking, prepare a jelly bag for straining the fruits. Scald the jelly bag by evenly pouring boiling water through, squeeze it well, then suspend over a large non-metallic bowl.

5 ▶ Ladle the fruit pulp into the jelly bag and leave to strain for a few hours, or overnight, but for not more than 24 hours. Make sure that the jelly bag is not squeezed or shaken while the pulp is straining, or the jelly will be cloudy.

6 ▲ Measure the strained juice. For each 600 ml juice, weigh 450 g sugar. Pour the juice into a clean preserving pan. Warm the sugar (see box, page 46), and add it to the juice. Stir over a low heat, with a wooden spoon, until the sugar has completely dissolved.

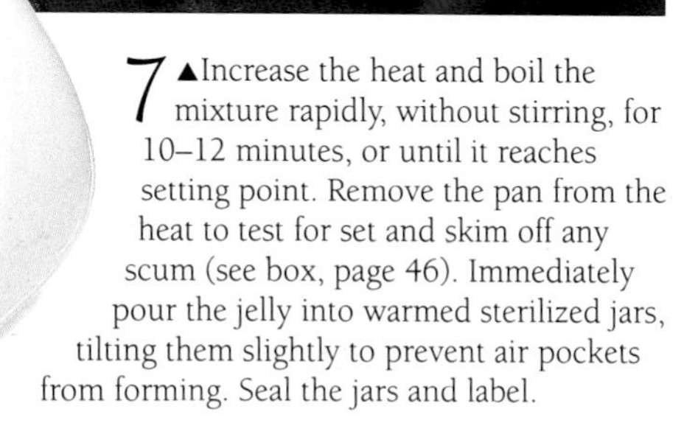

7 ▲Increase the heat and boil the mixture rapidly, without stirring, for 10–12 minutes, or until it reaches setting point. Remove the pan from the heat to test for set and skim off any scum (see box, page 46). Immediately pour the jelly into warmed sterilized jars, tilting them slightly to prevent air pockets from forming. Seal the jars and label.

APPLE AND ORANGE JELLY

INGREDIENTS

1.15 kg cooking apples

3 oranges

1.15 litres water

sugar

1 Peel, core, and chop the apples, reserving the cores and pips. Chop the unpeeled oranges and remove the pips. Put the apple cores and apple and orange pips on a square of muslin and tie up tightly into a bag with a long piece of string.

2 Put the chopped fruits with the water into a preserving pan and bring to a boil. Simmer for 50 minutes, or until the fruits are soft and pulpy. Remove and discard the muslin bag.

3 While the apples and oranges are cooking, prepare the jelly bag for straining the fruits. Scald it by evenly pouring boiling water through, squeeze it well, then use as directed (see box, page 41). Pour the fruit pulp into the jelly bag and leave to strain for a few hours or overnight, but for not more than 24 hours.

4 Measure the strained juice. For each 600 ml juice, weigh 450 g sugar. Pour the juice into a clean preserving pan. Warm the sugar (see box, left), and add it to the juice. Stir over a low heat until the sugar has completely dissolved.

5 Increase the heat and boil the mixture rapidly, without stirring, for 10–12 minutes, or until it reaches setting point. Test for set and skim off any scum (see box, page 46).

6 Pour the jelly into warmed sterilized jars, tilting them slightly to prevent air pockets from forming. Seal the jars and label.

Makes about 1.6 kg

WARMING SUGAR In these jelly recipes the amount of sugar required is calculated according to the amount of strained juice you have after the fruit is cooked. It is therefore difficult to be specific with an exact amount of sugar in the ingredients list, so make sure you have a good supply before you start. After weighing the sugar, it can be warmed so it dissolves quicker in the juice. Put the oven on its lowest setting, and put the sugar into an ovenproof bowl. Warm the sugar for about 15 minutes.

MIXED FRUIT JELLY

INGREDIENTS

450 g cooking apples

4 large oranges

1 lemon

1.75 litres water

450 g strawberries

sugar

1 Coarsely chop the apples, oranges, and lemon, without peeling, pipping, or coring them. Put them into a preserving pan with the water. Bring to a boil, and simmer for 40–45 minutes, or until the fruits are very soft and pulpy.

2 Hull and halve the strawberries and add them to the pan. Bring the mixture back to a boil, and simmer for 5 minutes, or until the strawberries are very soft.

3 While the strawberries are cooking, prepare the jelly bag for straining the fruits. Scald it by evenly pouring boiling water through, squeeze it well, then use as directed (see box, page 41). Pour the fruit pulp into the jelly bag and leave to strain for a few hours or overnight, but for not more than 24 hours.

4 Measure the strained juice. For each 600 ml juice, weigh 450 g sugar. Pour the juice into a clean preserving pan. Warm the sugar (see box, above), and add it to the juice. Stir over a low heat until the sugar has completely dissolved.

5 Increase the heat and boil the mixture rapidly, without stirring, for 30–35 minutes, or until it reaches setting point. Test for set and skim off any scum (see box, page 46). Immediately pour the jelly into warmed sterilized jars, tilting them slightly to prevent air pockets from forming. Seal the jars and label.

Makes about 1.5 kg

CHILLI AND APPLE JELLY

INGREDIENTS

1.4 kg cooking apples

600 ml water

300 ml cider vinegar

100 g fresh green chillies

sugar

2–3 drops of green food colouring (optional)

1 Chop the apples without peeling or coring them. Put them into a preserving pan with the water and vinegar.

2 Trim the chillies and cut in half lengthways. Remove and discard the seeds. Add the chillies to the pan with the apples.

3 Bring the mixture to a boil, and simmer for 20–25 minutes, or until the fruit is very soft and pulpy.

4 While the apples and chillies are cooking, prepare the jelly bag for straining the fruits. Scald it by evenly pouring boiling water through, squeeze it well, then use as directed (see box, page 41). Pour the pulp into the jelly bag and leave to strain for a few hours or overnight, but for not more than 24 hours.

5 Measure the strained juice. For each 600 ml juice, weigh 450 g sugar. Pour the juice into a clean preserving pan. Warm the sugar (see box, page 44) and add it to the juice. Stir over a low heat until the sugar has completely dissolved.

6 Increase the heat and boil the mixture rapidly, without stirring, for 8–10 minutes, or until it reaches setting point. Test for set and skim off any scum (see box, page 46).

7 Stir in the food colouring, if using. Immediately pour the jelly into warmed sterilized jars, tilting them slightly to prevent air pockets from forming. Seal the jars and label.

Makes about 450 g

COOK'S TIP
Chillies should be prepared with care. Wear rubber gloves when handling them, as it can be painful if any chilli gets on the skin, eyes, or nose. Afterwards, wash all surfaces and equipment which have come into contact with the chillies.

CHAMELEON
Despite using green grapes, Grape Jelly has a pink hue.

GRAPE JELLY

INGREDIENTS

1.4 kg green grapes

juice of 2 lemons

600 ml water

sugar

1 Coarsely chop the grapes. Put them into a preserving pan with the lemon juice and water. Bring the mixture to a boil, and simmer for 30 minutes, or until the fruit is very soft and pulpy.

2 While the grapes are cooking, prepare the jelly bag for straining the fruits. Scald it by evenly pouring boiling water through, squeeze it well, then use as directed (see box, page 41). Pour the fruit pulp into the jelly bag and leave to strain for a few hours or overnight, but for not more than 24 hours.

3 Measure the strained juice. For each 600 ml juice, weigh 450 g sugar. Pour the juice into a clean preserving pan. Warm the sugar (see box, page 44), and add it to the juice. Stir over a low heat until the sugar has completely dissolved.

4 Increase the heat and boil the mixture rapidly, without stirring, for 12 minutes, or until it reaches setting point. Test for set and skim off any scum (see box, page 46).

5 Immediately pour the jelly into warmed sterilized jars, tilting them slightly to prevent air pockets from forming. Seal the jars and label.

Makes about 675 g

Bramble Jelly

INGREDIENTS

2.25 kg blackberries

juice of 2 lemons

900 ml water

sugar

1 Put the blackberries into a preserving pan with the lemon juice and water. Bring to a boil, and simmer for 30 minutes, or until the fruit is very soft and pulpy.

2 While the blackberries are cooking, prepare the jelly bag for straining the fruit. Scald it by evenly pouring boiling water through, squeeze it well, then use as directed (see box, page 41). Pour the fruit pulp into the jelly bag and leave the blackberry pulp to strain for a few hours or overnight, but not more than 24 hours.

3 Measure the strained juice. For each 600 ml juice, weigh 450 g sugar. Pour the juice into a clean preserving pan. Warm the sugar (see box, page 44), and add it to the juice. Stir over a low heat, with a wooden spoon, until the sugar has completely dissolved.

4 Increase the heat and boil the mixture rapidly, without stirring, for 9–10 minutes, or until it reaches setting point. Test for set and skim off any scum (see box, left).

5 Immediately pour the jelly into warmed sterilized jars, tilting them slightly to prevent air pockets from forming. Seal the jars and label.

Makes about 1.4 kg

TESTING FOR SET AND SKIMMING JELLIES
Remove the pan from the heat. The best way to test for set is with a sugar thermometer, which should read 104°C (220°F) when setting point is reached. If you do not have a thermometer, use the cold plate test: drop a little of the jelly on a cold plate and chill quickly in the refrigerator. If it forms a skin and wrinkles when pushed with a finger, it should be ready. Lightly skim off any scum from the surface of the jelly, using a long-handled metal spoon.

Crab Apple Jelly

INGREDIENTS

2.75 kg crab apples

juice of 2 lemons

1.75 litres water

5 whole cloves

sugar

1 Coarsely chop the crab apples without peeling or coring them. Put them into a preserving pan with the lemon juice, water, and cloves. Bring the mixture to a boil, and simmer for 40 minutes, or until the fruit is very soft and pulpy.

2 While the apples are cooking, prepare the jelly bag for straining the fruit. Scald it by evenly pouring boiling water through, squeeze it well, then use as directed (see box, page 41). Pour the fruit pulp into the jelly bag and leave to strain for a few hours or overnight, but not more than 24 hours.

3 Measure the strained juice. For each 600 ml juice, weigh 450 g sugar. Pour the juice into a clean preserving pan. Warm the sugar (see box, page 44), and add it to the juice. Stir over a low heat until the sugar has completely dissolved.

4 Increase the heat and boil the mixture rapidly, without stirring, for 9–10 minutes, or until it reaches setting point. Test for set and skim off any scum (see box, above).

5 Immediately pour the jelly into warmed sterilized jars, tilting them slightly to prevent air pockets from forming. Seal the jars and label.

Makes about 1.8 kg

Fruit Butters & Cheeses

A GLUT OF sun-ripened fruits, available in the shops at bargain prices, means it is the time to make fruit butters and cheeses. They are so easy to prepare and require no special skill or equipment except a good wooden spoon for stirring. The bitter-sweet flavour and smooth spreadable texture of butters make them an ideal substitute for ordinary butter on bread, toast, scones, and cake. Fruit cheeses are more substantial, and they look good when set in shapely moulds, then turned out and sliced, cubed, or cut into wedges, whether sprinkled with sugar or not.

Although they have a very long history stretching back over thousands of years, fruit cheeses have never gone out of fashion or lost their popularity around the world. In the south of France, they are served with fresh cheeses; in Mexico and Brazil, guavas, quinces, mangoes, and pineapples are all used for making fruit cheeses, which are eaten with cream cheese and crackers. In England, orchard fruits, such as plums and damsons, are used to make colourful butters and cheeses.

Making Fruit Butters & Cheeses

❦ Choose the ingredients. Most fruits can be used, but sharply flavoured ones, such as blackberries, plums and quinces work best. Spices can be added. Granulated sugar is the recommended sweetener.

❦ Wash, stone or hull, and chop the fruits as necessary, removing any mouldy or damaged parts. Put the fruits into a preserving pan. (Lightly greasing the pan first helps prevent the fruits from sticking.) Barely cover the fruits with water and bring to a boil, then simmer, stirring occasionally over a low heat, until the fruits are soft and pulpy and there is no excess liquid. Sieve or liquidize the fruits to a purée.

❦ Weigh the pulp. For butters, weigh 225–350 g sugar for each 450 g pulp. For cheeses, weigh 450 g sugar for each 450 g pulp. Warm the sugar (see box, page 44). Return the pulp to the pan with the sugar, add spices if using, and stir over a low heat until the sugar has completely dissolved.

❦ Bring to a boil and simmer, 30–45 minutes for butters, and 45–60 minutes for cheeses, stirring frequently to prevent them from burning. The butter is ready when a spoonful placed on a saucer does not exude any liquid. The cheese is ready when a spoon drawn across the bottom of the pan leaves a clean line.

❦ Spoon butters into sterilized jars (see page 11), and seal. For cheeses, pour the mixture into moulds, from which they can be turned out for slicing. Brush the moulds with a little glycerine, so the cheeses will slip out easily.

How to seal and store

Screw-top lids are the best choice for sealing fruit butters, because butters have a tendency to dry out, and need an airtight seal. Butters, which have a thick spreading consistency, do not keep quite as well as jams and should be used up within 3 months. If cheeses are in moulds, cover them as you would jams, with wax discs and cellophane circles (see page 15). Cheeses have a higher sugar content than butters and will keep for up to 1 year.

What can go wrong and why

If the butter starts to ferment, this could be a result either of too short a cooking time or of too little sugar. The mixtures may caramelize if the excess water has not evaporated before adding the sugar.

STRAWBERRY-PEAR BUTTER

The flavours of summer and autumn blend together in this thick, creamy butter. Serve it in summer spooned on to freshly made pancakes with sliced strawberries and cream, or on a winter's afternoon spread on hot buttered scones or toast. Small jars of this fruit butter also make pretty gifts for special occasions.

INGREDIENTS

1.8 kg pears

450 g strawberries

300 ml water

1/2 tsp ground cinnamon

sugar

Makes about 775 g

1 ▼Peel the pears, cut into quarters, and remove and discard the cores. Chop the pears finely. Hull the strawberries and cut them into quarters.

2 Put the pears and strawberries into a preserving pan. Add the water and ground cinnamon and bring to a boil. Simmer, stirring occasionally, for 1–1 1/4 hours, or until the fruits are soft. The mixture should be thick and pulpy, and all of the excess liquid should have evaporated.

3 ◄With the back of a wooden spoon, press the fruit mixture through a plastic sieve set over a non-metallic bowl. Weigh the sieved pulp. For each 450 g pulp, weigh 350 g sugar. Warm the sugar (see box, page 44).

4 Return the pulp to the pan, add the sugar, and stir the mixture over low heat, with a wooden spoon, until the sugar has completely dissolved.

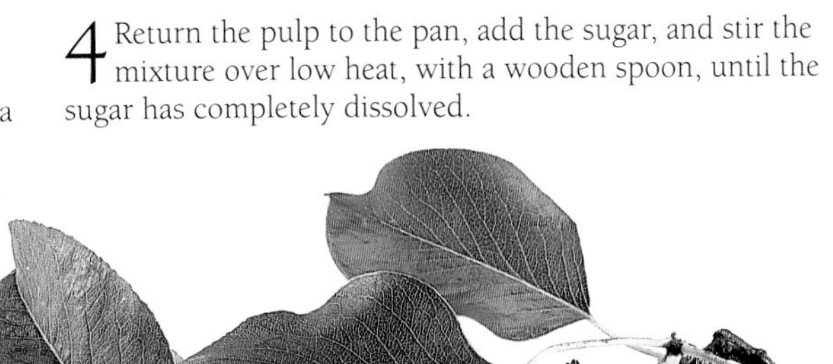

5 ▼Simmer over low heat, stirring frequently, until the butter thickens to resemble clotted cream. This will take 15–20 minutes. The butter is ready when a spoonful placed on a saucer does not exude any liquid.

6 Remove the pan from the heat. Spoon the butter into warmed sterilized jars, to within 3 mm of the tops. Seal the jars and label.

PEACH BUTTER

INGREDIENTS

1 kg ripe peaches

480 ml water

sugar

1 Drop the peaches into boiling water for 1–2 minutes, then transfer to a bowl of iced water. Drain the peaches when cool.

2 Halve the peaches, peel, and coarsely chop the flesh. Put the flesh into a saucepan with the measured water. Simmer, stirring occasionally, for 20 minutes, or until very soft. Press the mixture through a plastic sieve set over a non-metallic bowl.

3 Weigh the sieved pulp. For each 225 g pulp, weigh 100 g sugar. Warm the sugar (see box, page 44). Return the pulp to the pan, add the sugar, and stir over a low heat until the sugar has completely dissolved. Simmer over a low heat, stirring frequently, for 30 minutes, or until the butter thickens to resemble clotted cream.

4 Spoon the butter into warmed sterilized jars, seal, and label.

Makes about 450 g

CRANBERRY BUTTER

INGREDIENTS

1.4 kg frozen cranberries

300 ml water

sugar

1 Put the cranberries into a saucepan with the water and bring to a boil. Simmer, stirring occasionally, for 35–40 minutes, or until the cranberries are very soft. Press the mixture through a plastic sieve set over a non-metallic bowl.

2 Weigh the sieved pulp. For each 450 g pulp, weigh 350 g sugar. Warm the sugar (see box, page 44). Return the pulp to the pan, add the sugar, and stir over a low heat until the sugar has completely dissolved.

3 Simmer over a low heat, stirring frequently, for 1 hour, or until the butter thickens to resemble clotted cream.

4 Spoon the butter into warmed sterilized jars, seal, and label.

Makes about 1.15 kg

CIDER-APPLE BUTTER

INGREDIENTS

1 kg eating apples

1 kg cooking apples

240 ml cider

sugar

1 tsp ground coriander

1 Coarsely chop the apples, without peeling or coring them. Put the apples into a saucepan, add the cider, and bring to a boil. Simmer, stirring occasionally, for 25–30 minutes, or until the apples are very soft. Press the mixture through a plastic sieve set over a non-metallic bowl.

2 Weigh the sieved pulp. For each 100 g pulp, weigh 50 g sugar. Warm the sugar (see box, page 44). Return the pulp to the pan, add the sugar and coriander, and stir over a low heat until the sugar has completely dissolved.

3 Simmer over a low heat, stirring frequently, for 30 minutes, or until the butter thickens to resemble clotted cream.

4 Spoon the butter into warmed sterilized jars, seal, and label.

Makes about 675 g

VARIATION

Give a slightly tangier taste to this all-time favourite butter by using Spiced Pickling Vinegar (see recipe, page 120) in place of the cider, and ground cinnamon instead of the coriander. Add the finely grated zest of 1 lemon with the spice.

PLUM CHEESE

INGREDIENTS

2.75 kg sweet green plums

300 ml water

sugar, warmed (see box, page 44)

1 Put the plums and water into a preserving pan and bring to a boil. Simmer, stirring occasionally, for 30 minutes, or until tender.

2 Press the plums and juice through a plastic sieve set over a non-metallic bowl. Discard the stones. Weigh the sieved pulp and return it to the pan with an equal weight of warmed sugar. Stir over a low heat until the sugar has completely dissolved.

3 Bring to a boil and simmer, stirring frequently, for 40–45 minutes. The pulp should be thick and firm enough that when a wooden spoon is drawn across the bottom of the pan, it will not move.

4 Spoon the cheese into warmed sterilized jars, to within 3 mm of the tops. Seal and label.

Makes about 1.8 kg

GUAVA CHEESE

INGREDIENTS

450 g guavas

100 ml water

225 g sugar, warmed (see box, page 44)

1 1/2 tbsp lemon juice

This cheese is a quick, no-fuss preserve with lots of tropical flavour.

1 Chop the guavas coarsely and put with the liquid into a food processor. Purée the mixture until it is smooth.

2 Put the purée, warmed sugar, and lemon juice into a saucepan. Stir over a low heat until the sugar has completely dissolved. Simmer over a low heat for 12 minutes. The pulp should be thick and firm enough that when a wooden spoon is drawn across the bottom of the pan, it will not move.

3 Spoon the cheese into warmed sterilized jars, to within 3 mm of the tops. Seal and label.

Makes about 350 g

DID YOU KNOW?
Guavas are an oval tropical fruit with a yellow to green skin and an off-white to red flesh. To prepare, cut the fruit in half and scoop out the soft flesh and seeds. Discard the skin.

PLUM AND LEMON CHEESE

INGREDIENTS

2.75 kg plums

300 ml water

sugar, warmed (see box, page 44)

finely grated zest and juice of 2 lemons

1 Put the plums and water into a preserving pan and bring to a boil. Simmer, stirring occasionally, for 30 minutes.

2 Press the plums and juice through a plastic sieve set over a non-metallic bowl. Discard the stones. Weigh the sieved pulp and return it to the pan with an equal weight of warmed sugar. Add the lemon zest and juice. Stir over a low heat until the sugar has completely dissolved.

3 Bring to a boil, then simmer over a low heat, stirring frequently, for 40–45 minutes. The pulp should be thick and firm enough that when a wooden spoon is drawn across the bottom of the pan, it will not move.

4 Spoon the cheese into warmed sterilized jars, to within 3 mm of the tops. Seal and label.

Makes about 1.8 kg

English Fruit Curds

Lemon curd tarts, with their sharp, fruity flavour, are what the English think of when curds are mentioned, unless of course their minds stray back to childhood and Little Miss Muffet, who ate curds and whey until she was disturbed by a spider. These curd recipes, unlike Miss Muffet's curds, are enriched with butter and eggs. Although lemon curd is an undoubted favourite, there are other, equally delicate and delicious, fruit curds to beguile our palates and to take as gifts to friends. Oranges, limes, and other fruits can be used, and there is plenty of room for experimentation.

Curds have the paradoxical virtue of not being good keepers. They must be refrigerated, and should not be left to languish on the refrigerator shelf, but used within a month. In any case, these preserves are better eaten sooner than later because their flavour does fade and their fresh quality is lost the longer they are kept.

Curds can be spread on bread, toast, or on plain cake. They can also form the base of a dessert, or be spooned over ice-cream with colourful pieces of fruit. And what better way to spend a miserably wet Sunday afternoon in winter than making a batch of curd tarts, baked in shortcrust pastry that melts in the mouth, which recreate teatimes of the past?

Making Fruit Curds

- Select fresh ingredients, particularly fresh eggs. Caster sugar is the preferred sweetener, because curds are cooked very gently and the fine granules will dissolve more easily.
- Finely grate the zest of the chosen citrus fruits and squeeze out the juice. Strain and measure the juice. Pour it into a non-metallic bowl set over a pan of gently simmering water (or use a double boiler), and add the grated zest, butter, and sugar. Stir the mixture over a low heat, with a wooden spoon, until the butter has melted and the sugar has completely dissolved. Lightly beat the eggs and strain into the mixture. Continue simmering, stirring, until the mixture has thickened enough to coat the back of the spoon. Do not allow the contents of the bowl to boil. Curds require long slow cooking, so do not rush the process by increasing the heat.
- When the mixture has thickened, immediately pour it into warmed sterilized jars (see page 11). Because of their limited shelf life, curds are best made in small quantities and put into 225 g jars. Fill them right to the tops, because the curd will shrink and thicken as it cools.

How to seal and store

Use wax and cellophane circles, because they help prevent curds from going mouldy. Place a wax circle, wax-side down, on the top of the curd, smooth over to remove any air pockets, and leave to cool. Put a dampened cellophane circle over the top, and secure with an elastic band. Label and keep in the refrigerator. Use within 1 month.

What can go wrong and why

If a curd is runny, it has been insufficiently cooked. If it is curdled, it was cooked over too high a heat.

Cook's Tip It is important to thicken a curd very slowly, without boiling, to avoid curdling. If this does occur, remove the curd from the heat immediately and whisk vigorously.

LEMON CURD

The tang of this spread comes from the lively lemon, and the thick, luscious smoothness from the careful blending of fresh eggs, creamy butter, and fine-grained sugar. Lemon curd will only keep for a month in the refrigerator, but it is so delicious that it is unlikely to last this long anyway. Serve it on toast, in tarts, or as a filling for cakes.

INGREDIENTS

6–8 large juicy lemons

225 g unsalted butter

575 g caster sugar

5 large eggs

Makes about 1.15 kg

WAXED LEMONS
If the lemons are waxed, scrub them to remove the coating before grating the zest.

1 ▶ Grate the zest from the lemons on the finest side of the grater. Squeeze the juice and strain it into a large measuring jug. You will need 300 ml juice.

2 ▼ Cut the butter into small pieces. Put the pieces of butter into a glass bowl, along with the sugar, lemon zest and juice, set over a pan of gently simmering water. The bottom of the bowl should not touch the water, nor should the water boil rapidly. Stir the mixture until the butter has melted and the sugar has completely dissolved.

3 ▼ Lightly beat the eggs in a bowl, but do not whisk them. Strain the eggs through a plastic sieve into the lemon mixture. Simmer over a low heat, stirring constantly with a wooden spoon, until the mixture thickens slightly. This will take 20–25 minutes. Do not allow the mixture to boil or it will curdle.

4 ▲ As soon as the mixture is thick enough to coat the back of the spoon, remove the bowl from the pan of water.

5 ▲ Pour into warmed sterilized jars. Place a circle of waxed paper, wax side down, on top of each jar. Smooth over to remove air pockets. Leave to cool.

6 ▲ Cover with dampened cellophane circles. Label, and store the curd in the refrigerator.

More Curd Flavours

Oranges, limes, and grapefruit all make delicious curds. By adapting the Lemon Curd recipe you can create a sweeter or tangier preserve. For a touch of sophistication, add a splash or two of a flavoured liqueur.

Above, Lime Curd; right, Orange Curd

Orange Curd

This curd is delicious served with warm pancakes and whipped cream, particularly if you squeeze extra orange juice on top. Follow the recipe for Lemon Curd, using the juice of 3 medium oranges and the juice of 1 lemon instead of the juice of 6–8 lemons. Combine the squeezed juices from the oranges and lemon to obtain a total of 300 ml.

Lime Curd

Use limes to make a tangier spread; the juice and tiny flecks of grated zest will turn the curd a subtle shade of green. Lime Curd makes an unusual filling for a sponge cake, or a tasty topping for ice cream. You will need more limes than lemons for this curd, about 10 juicy ones, to get the 300 ml juice required.

Pink Grapefruit and Lime Curd

◀ Use the finely grated zest of 2 pink grapefruit and 1 lime for this curd, and combine the juices to ensure you have the amount you need. If the curd is too runny, thicken it by sprinkling a little ground rice into the mixture, stirring it in well.

Tipsy Curd

Stir a few spoonfuls of a rich liqueur, such as advocaat, into the finished curd. Tipsy curd is delicious when spread generously in a pastry pie shell, then piled high with sliced tropical fruits.

Did You Know?

Tangy fruit curds, simply spread on bread or between layers of sponge cake, were popular teatime treats in Victorian and Edwardian England. They were also used as a fragrant filling for trifles, and piped into tartlets in elegant sweet swirls.

In her book *Modern Cookery for Private Families*, the nineteenth-century cookery writer Eliza Acton used a type of lemon curd as a filling for little puff pastry tarts. In addition to eggs, sugar, pounded butter, and lemon, she instructed the reader to "strew lightly in a spoonful of flour, well dried and sifted". Most modern fruit curds do not include flour, although the idea of adding a thickener is a good one if the curd seems a little on the runny side – ground rice is suggested in Pink Grapefruit and Lime Curd for this reason.

FRUITS IN
ALCOHOL
& FLAVOURED
DRINKS

Fruits in Alcohol

Nothing is quite so luxurious as fruit that has soaked up the perfumes of fortified wine, spirits, or liqueurs and, in turn, given its own flavour to the macerating liquid. Something quite astonishing happens to everyday pitted prunes when they have spent some time immersed in port.

Rum, gin, and brandy all have a strong affinity for summer and autumn fruits, but almost any kind of alcohol can be used. Kirsch, for example, complements full-flavoured fruits, such as raspberries, blackberries, loganberries, and pineapples; whisky steeps well with grapes, and apricots take on a wonderful new flavour after being soaked in orange curaçao. Some cooked fruits can be preserved in wine, but they will last longer if used in combination with a spirit.

Fruits in alcohol, sweetened with sugar, look beautiful, taste even better, and are easy to make with only a little time and effort. They are superb when served as desserts, especially when topped with whipped cream, and their richness makes them great accompaniments to roast meats, game, and poultry, especially duck.

Making Fruits in Alcohol

- Choose your fruits; they must be just ripe. If they are over-ripe they will not retain their shape in the alcohol. Pick over the fruits, removing any mouldy or damaged parts. Wash and dry the fruits and remove stones, if necessary. Remove and discard any excess peel, pips, and cores from fruits such as pears. Soft fruits, such as berries, blackcurrants, apricots, and peaches, can be used raw, while firmer ones, such as apples and hard plums, may require light cooking first.
- Pack the fruits into sterilized jars (see page 11), either in layers with sugar or in a sugar syrup. Warm the jars if the ingredients used are hot. White granulated sugar is the usual choice of sweetener, but brown sugar can be used for flavour, or if a darker colour is preferred. Shake the jars once or twice during the early days of storage, to help dissolve the sugar (unless a syrup is used) and blend it with the alcohol.
- Pour the chosen alcohol over the fruits to completely cover them, making sure that there are no air pockets between them. The amount of liquid can vary, depending on the size of the fruits and the amount of sugar used, so make sure you have plenty of the chosen alcohol in stock.
- Spices, such as whole cloves, cinnamon sticks, and allspice berries can be added to the fruits for extra flavour.

How to seal and store

Seal tightly, preferably with non-corrosive, screw-top lids. An airtight seal is required to prevent the alcohol from evaporating. Leave the fruits in a cool dark place for at least 1 month, and preferably longer, before using, to allow the flavours to develop. Shake the jars from time to time to blend the sugar and flavourings. As long as the fruits are covered by the alcohol, they should keep for about 1 year, and the flavour will improve over this time.

What can go wrong and why

Some fruits have a tendency to float to the top of the liquid. Prevent this either by filling the jar with fruits to the very top, or by placing a piece of crumpled greaseproof paper in the top of the jar to hold the fruits down. Remove the paper after about 1 week, after which time the fruits should remain in place. Make sure the fruits are covered with alcohol before re-sealing the jar. If the alcohol has evaporated, and left the fruits exposed, the jar has not been sealed tightly enough.

SUMMER FRUITS IN KIRSCH

A delicious way to preserve soft fruits, this recipe can be made either all at once, or throughout the summer as the different fruits ripen. If you layer the fruit over a period of time, keep it below the level of the alcohol by putting a lightly weighted small saucer or some crumpled greaseproof paper in the jar. When the last fruits have been added, top up with kirsch, and re-seal.

1 ◀Pick over the fruits, removing any leaves, mouldy or damaged parts, hulls, and stalks. For the currants, run a fork along the stalks to remove the berries. Wash the fruits in a plastic sieve or colander and dry on kitchen paper.

INGREDIENTS

675 g mixed summer fruits, e.g. raspberries, boysenberries, blueberries, small strawberries, blackberries, redcurrants, whitecurrants, and blackcurrants

250 g sugar

about 300 ml kirsch

Fills one 1 litre jar

2 ▶ Layer the fruits and sugar in a sterilized jar to within 2.5 cm of the top of the jar.

3 ▲Pour in the kirsch to cover the fruits by about 1.25 cm, making sure there are no air pockets left between the fruits. Seal the jar and label.

4 Keep in a cool dark place for 2–3 months before using, to allow the flavours to develop, tipping the jar occasionally so the sugar dissolves.

TWICE THE FUN
Alcohol-drenched fruits make an intoxicating dessert, while the berry-flavoured kirsch can be drunk separately.

Mangoes in Brandy

Ingredients

3–4 large mangoes

225 g sugar

450 ml brandy

1 Cut each mango on both sides of the stone, slice the flesh away from the stone, and discard the stone. Peel the mango slices and cut the flesh into even-sized chunks. There should be about 1 kg mango chunks.

2 Pack the mango chunks into sterilized jars, layering them with the sugar as you go. Leave 2.5 cm at the tops of the jars.

3 Pour in the brandy to cover the mango by 1.25 cm, making sure there are no air pockets between the fruits. Seal the jars, label, and shake well.

4 Keep in a cool dark place for at least 2 months before using, to allow the flavours to develop. Shake from time to time during the first week of storage, to make sure all the sugar dissolves in the brandy.

Fills about three 480 ml jars

Pears in Vodka

Ingredients

1 kg ripe pears

450 g sugar

600–750 ml vodka

1 Peel, quarter, and core the pears quickly to prevent discoloration. Pack the pears into sterilized jars, layering them with the sugar as you go. Leave 2.5 cm at the tops of the jars.

2 Pour in the vodka to cover the pears by 1.25 cm, making sure there are no air pockets between the pear quarters. Seal the jars, label, and shake well.

3 Keep in a cool dark place for at least 2–3 months before using, to allow the flavours to develop. Shake from time to time during the first week of storage, to make sure all the sugar dissolves.

Fills about two 750 ml jars

Plums in Rum

Ingredients

1.4 kg plums

225 g sugar

100 ml water

350 ml white rum

Rumtopf Use a mixture of fruits, such as cherries, strawberries, apricots, nectarines, and plums, to make a traditional *Rumtopf*. This German preserve is made in stages as the fruits become available, to keep until Advent. Layer each addition of fruit with sugar and cover with rum as directed.

1 Prick the plums all over with a sterilized needle or a wooden toothpick. In a saucepan, combine the sugar and water and simmer over a low heat, stirring with a wooden spoon, until the sugar has completely dissolved. Add the plums and simmer for 5 minutes. Remove the pan from the heat and allow to cool.

2 Pack the plums and syrup into sterilized jars, then pour in the rum to cover them by 2.5 cm. Seal the jars and label.

3 Keep in a cool dark place for at least 1 month before using, to allow the flavours to develop. Check after the first day or so that the plums are still covered by the liquid; add more rum, if necessary and re-seal.

Fills about four 600 ml jars

Prunes in Port

Ingredients

500 g stoned prunes

480–600 ml tawny port

Serve the prunes with roast pork or game, or as a simple dessert with ice cream, spooning the port over it. Or drink the port as a liqueur.

1 Pack the prunes into a sterilized jar and pour in about 480 ml port to cover the prunes by 4 cm. Cover and reserve the remaining port.

2 Seal the jar and label. Keep the prunes in a cool dark place.

3 Check the prunes in a day or two. Once they have soaked up so much port that they are no longer covered, pour in the remaining port. Re-seal and leave for 1 month before using, to allow the flavours to develop.

Fills one 1 litre jar

Tipsy Apricots

Ingredients

675 g–1 kg apricots, according to size

350 g sugar

240–300 ml orange curaçao

240–300 ml dark rum

1 Prick the apricots all over with a sterilized needle or a wooden toothpick.

2 Pack the apricots into sterilized jars, layering them with the sugar as you go. Leave 2.5 cm at the tops of the jars.

3 Mix the orange curaçao with the rum in a glass measuring jug. Pour the mixed alcohols into the jars to cover the apricots by 1.25 cm, making sure there are no air pockets between the fruits.

4 Seal the jars and label. Shake the jars well so that the sugar can start to blend and dissolve into the alcohol.

5 Keep in a cool dark place for at least 2 months before using, to allow the flavours to develop. Shake from time to time during the first week to of storage, to make sure all the sugar dissolves in the alcohol.

Fills about two 750 ml jars

Variation
Mandarins can be preserved in the same syrup as apricots. Peel the fruits, keeping them whole (there is no need to prick them). Pack the fruits into jars, layering with the sugar and the grated zest of 2 oranges, and continue as directed in step 3.

Pears in Brandy

Ingredients

1.4 kg firm ripe pears

100 g sugar

480 ml brandy

1 Peel, quarter, and core the pears. If the fruits are small, just halve and core them.

2 Put the pear quarters and the sugar into a saucepan with just enough water to cover. Cover the pan and simmer for 30 minutes, or until the pears are tender and the sugar has completely dissolved. Allow the pears to cool, then transfer into sterilized jars, using a slotted spoon.

3 Boil the pear juice over a high heat until it has thickened and reduced to 240 ml. Pour it over the pears. Pour in the brandy to within 2.5 cm of the tops.

4 Seal the jars and label. Keep in a cool dark place for at least 1 month before using, to allow the flavours to develop.

Fills about three 300 ml jars

Cook's Tip To prevent discoloration, cover the peeled pears with lemon juice, or place in acidulated salted water made with 1 tsp salt and 1 tsp citric acid to each 1 litre water.

GRAPES IN WHISKY

INGREDIENTS

1–1.1 kg seedless grapes

450 g sugar

600–750 ml whisky

1 Prick the grapes all over with a sterilized needle or a wooden toothpick. Pack the grapes into sterilized jars, layering them with the sugar as you go. Leave about 2.5 cm space at the tops of the jars.

2 Slowly pour in the whisky to cover the grapes by 1.25 cm, making sure that there are no air pockets.

3 Seal the jars, label, and shake well. Keep in a cool dark place for at least 2–3 months before using, to allow the flavours to develop. Shake from time to time during the first week of storage to make sure that all the sugar dissolves.

Fills about two 750 ml jars

PEARS IN RED WINE

INGREDIENTS

600 ml red wine

450 g sugar

2 kg pears

2 cinnamon sticks

brandy

1 Put the red wine and sugar into a preserving pan and stir over a low heat until the sugar has completely dissolved.

2 Peel, quarter, and core the pears, and add to the pan with the cinnamon sticks. Bring the mixture to a boil, lower the heat, and simmer the pears for 5–10 minutes, or until they are just tender. Take care not to overcook. Discard the cinnamon sticks. With a slotted spoon, transfer the pears to sterilized jars, to within 2.5 cm of the tops of the jars.

3 Increase the heat and boil the syrup rapidly, without stirring, for 5 minutes. Strain the syrup into a measuring jug and make up to 600–750 ml with brandy. Pour the mixture into the jars to cover the pears by 1.25 cm. Seal the jars and label. Keep in a cool dark place for at least 2–3 months before using, to allow the flavours to develop.

Fills about two 750 ml jars

VARIATION The pears can be replaced by mixed soft red fruits, such as strawberries, raspberries, and redcurrants. Pack, uncooked, into sterilized jars. Prepare the red wine syrup as directed in step 1, adding 6 crushed cardamom pods. Strain the syrup and pour over the red fruits to cover.

PAWPAWS IN RUM

INGREDIENTS

3 pawpaws

50 g pistachios

175 g sugar

325 ml rum

1 Peel and halve the pawpaws. Scoop out the seeds. Cut the flesh into large cubes. There should be about 575 g fruit.

2 Put the pistachios into a bowl and pour over boiling water to cover. Leave for 1 minute, then drain and transfer to a bowl of cold water. Drain again, then rub off the skins with your fingers.

3 Pack the pawpaws and pistachios into sterilized jars, layering them with the sugar as you go. Leave about 2.5 cm at the tops of the jars. Pour in the rum to cover the fruit and nuts by 1.25 cm, making sure that there are no air pockets.

4 Seal the jars, label, and shake well. Keep in a cool dark place for at least 2–3 months before using, to allow the flavours to develop. Shake from time to time during the first week of storage to make sure that all the sugar dissolves.

Fills about two 600 ml jars

PINEAPPLES IN KIRSCH

INGREDIENTS

2 ripe pineapples, total weight about 2 kg

450 g sugar

450–600 ml kirsch

Kirsch is a cherry brandy which has a natural affinity with pineapple.

1 Slice the tops off the pineapples. Cut the peel off in strips, cutting deep enough to remove the "eyes" with the peel. Cut the pineapple across into thick slices, and cut out and discard the hard central core from each slice. Dice the flesh.

2 Pack the pineapple into sterilized jars, layering it with the sugar as you go. Leave about 2.5 cm space at the tops.

3 Pour in the kirsch to cover the pineapple by 1.25 cm, making sure that there are no air pockets between the fruits. Seal the jars, label, and shake well.

4 Keep in a cool dark place for at least 1–2 months before using, to allow the flavours to develop. Shake from time to time during the first week of storage to make sure that all the sugar dissolves.

Fills about two 600 ml jars

RAISINS IN GENEVER WITH JUNIPER BERRIES

INGREDIENTS

675 g raisins

350 g sugar

3 tsp juniper berries

750–900 ml genever gin

Gin is a spirit that is made from distilling juniper berries. It is therefore appropriate for gin and juniper to be mixed together in this recipe.

1 Layer the raisins with the sugar in sterilized jars, adding a few juniper berries with each layer.

2 Pour in the genever gin to cover the raisins, filling to the tops of the jars and making sure that there are no air pockets between the raisins. Seal the jars, label, and shake well.

3 The next day, check to see that the raisins have not risen above the level of the alcohol. If they have, remove a layer, and re-seal. Keep in a cool dark place for at least 2 months before using, to allow the flavours to develop. Shake from time to time during the first week of storage to make sure that all the sugar dissolves.

Fills about two 750 ml jars

DID YOU KNOW?

Genever is a type of Dutch gin with a strong flavour. In Holland, raisins in genever are traditionally served in glasses on New Year's Eve. If you cannot get genever, ordinary gin can be used.

PEACHES IN BRANDY

INGREDIENTS

675 g–1 kg peaches, depending on size

350 g sugar

450–480 ml brandy

1 Halve and stone the peaches. The fruits can also be peeled for a classic effect, but this is not necessary.

2 Pack the peach halves into sterilized jars, layering with the sugar as you go. Leave at least 2.5 cm at the tops of the jars.

3 Pour in the brandy to cover the peaches by 1.25 cm, making sure that there are no air pockets between the fruits. Seal the jars, label, and shake well.

4 Keep in a cool dark place for at least 2–3 months before using, to allow the flavours to develop. Shake from time to time during the first week of storage to make sure that all the sugar dissolves.

Fills about two 900 ml jars

Flavoured Wines & Spirits

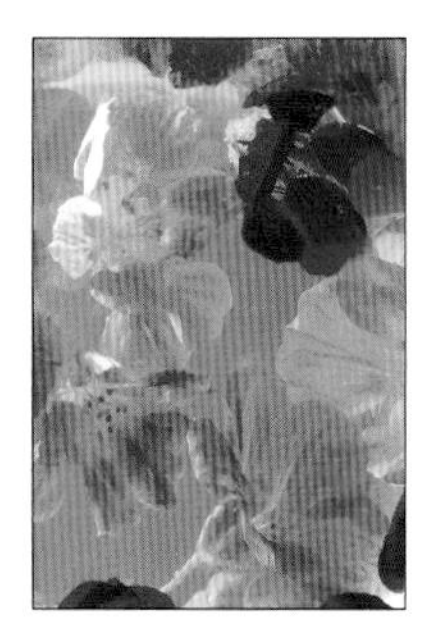

Almost all commercially flavoured wines and spirits are made according to jealously guarded formulae. Aquavit with caraway seeds, gin with juniper berries, sambuca and ouzo with aniseed, to name but a few. Fruit brandies, known as *alcools blancs*, are commercially distilled from fresh fruits, the most popular being plums, Williams pears, black cherries, and raspberries, while the production of vermouths requires a complicated process involving herbs, sugar syrup, alcohol, and all manner of equipment.

Yet flavouring wines and spirits is not altogether out of reach of the home cook. Vanilla and cherry brandies are easy to make, while hot red chillies can transform the taste of sherry. Dry white wine is delicious when delicately flavoured with fruit, with the aid of a little sugar syrup. All it requires is enough patience to wait while the fruit and the wine exchange flavours. Although the wine will be the winner, the strained-out fruit can often be used, perhaps as an ingredient in a home-made ice-cream or sorbet.

Making Flavoured Wines & Spirits

❦ Sterilize a large glass jar or bottle (see page 11) in which to steep the flavourings and alcohol.

❦ Choose the ingredients. Fresh berries and stone fruits are the best choice of fruits, and they should be ripe, but not over-ripe. Herbs and spices and other seasonings should be as fresh as possible. If flavouring with spice, use whole spices. The choice of alcohol depends on personal preference – any kind can be used, provided that it has an alcoholic content of at least 37.5% by volume.

❦ If fruits are being used as a flavouring, wash and dry them first. Remove any stalks, and any mouldy or damaged parts. Halve and stone fruits such as apricots. Cherries do not require stoning, but should be pricked all over with a sterilized needle or a wooden toothpick so that their flavour and colour can be released.

❦ Put the fruits into the prepared jar, layering it with sugar or pouring a sugar syrup over it, then add the alcohol. For herb- and spice-infused wines and spirits, simply insert the flavourings into sterilized bottles (lightly bruising fresh herbs first), and pour over the chosen alcohol.

❦ Allow the flavours to develop before using, shaking the bottles or jars from time to time. Allow about 3 months for fruit-infused spirits. Wines should be left to steep for 2–3 days if combined with fresh ingredients, and for up to 4 weeks if dried flavourings, such as apricots, herbs, and spices, are used.

❦ Strain the flavoured wine or spirit through a double layer of muslin into sterilized bottles.

How to seal and store

Seal the bottles tightly with non-corrosive screw-top lids or corks. Choose new corks, sterilize them, and secure into the necks of the bottles with a wooden mallet. Spirits keep well and last almost indefinitely when stored in a cool dark place. Flavoured wines do not last as long. Once opened, the flavour will deteriorate after a few days, unless the wine has been combined with other preservatives (sugar and spirits).

What can go wrong and why

A flavoured wine or spirit will go cloudy if ground spices are used for flavouring. Always use whole spices. If the drink lacks flavour, the infusing time has been too short, or the ingredients are too bland.

CHERRY BRANDY

Traditionally, Morello cherries are used for this classic fruited spirit because they are less sweet than other varieties, but you can use any dark, flavoursome cherry instead. Let the brandy mature before decanting it into bottles, in time to give as a cheering winter gift. And don't neglect the cherries. After straining, they are delicious topped with fresh whipped cream.

INGREDIENTS

450 g cherries

75 g sugar

2 drops of almond essence

600 ml brandy

Fills one 1 litre bottle

1 ◀ Remove all the cherry stalks. Prick each cherry all over with a sterilized needle or a wooden cocktail stick.

2 ◀ Layer the cherries with the sugar in a large sterilized jar, to within 2.5 cm of the top. Add the almond essence to the jar.

3 ▶ Pour in the brandy to cover the cherries by 1.25 cm. Seal the jar and shake well. Keep in a cool dark place for at least 3 months before using, to allow the flavours to develop. Shake the jar from time to time.

4 ▶ Line a funnel with a double layer of muslin and strain the brandy through it into a sterilized bottle. Seal the bottle and label. The brandy is now ready to use.

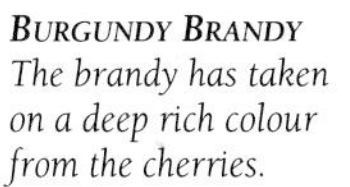

BURGUNDY BRANDY
The brandy has taken on a deep rich colour from the cherries.

Plum Gin

INGREDIENTS

450 g plums

225 g sugar

1.15 litres gin

a few drops of almond essence (optional)

1 Remove any stalks from the plums. Prick each plum all over with a sterilized needle or a wooden toothpick.

2 Layer the plums and sugar in a sterilized jar. Pour in the gin to cover the plums completely and add the almond essence, if using. Seal the jar and shake well. Keep in a cool dark place for about 3 months before using, to allow the flavours to develop. Shake the jar from time to time.

3 Line a funnel with a double layer of muslin and strain the flavoured gin through it into sterilized bottles. Seal the bottles and label. The gin is now ready to use.

Makes about 1.15 litres

Apricot Wine with Brandy

INGREDIENTS

100 g dried apricots

75 g sugar

1 bottle dry white wine

1 1/2 tbsp brandy

1 Coarsely chop the apricots and put them into a saucepan with the sugar. Pour in just enough water to cover. Cover the pan and leave to stand for 2 hours. Simmer the apricots, stirring occasionally, for 10 minutes, or until the fruit is soft and tender.

2 Put the apricots and their syrup into a sterilized jar and slowly pour in the white wine and brandy to cover. Seal the jar and shake well, to remove any air pockets. Keep in a cool dark place for 3 weeks before using, to allow the flavours to develop. Shake the jar from time to time.

3 Line a funnel with a double layer of muslin and strain the flavoured wine through it into sterilized bottles. Seal the bottles and label. The wine is now ready to use.

Makes about 900 ml

COOK'S TIP Serve Apricot Wine with Brandy well chilled, as an aperitif. For a special occasion, pour it into a punch bowl and float nasturtium flowers on the top. The strained apricots can be served separately as a dessert, or as an accompaniment to savoury game, meat, or poultry dishes.

Blushing Strawberry Wine

INGREDIENTS

225 g strawberries

3 sprigs of fresh lemon balm

1 sprig of fresh rosemary

1 bottle of semi-sweet rosé wine

This is a refreshing early summer drink which can be served either undiluted, or in tall glasses with a splash of sparkling water. Float vivid blue borage flowers or a sprig of lemon balm in each glass, for an attractive finish.

1 Hull and slice the strawberries. Lightly bruise the herbs to release their flavours. Put the strawberries and herbs into a sterilized jar and pour in the wine.

2 Seal the jar and shake well. Keep in a cool dark place for 2 days before using, to allow the flavours to develop. Shake the jar from time to time.

3 Line a funnel with a double layer of muslin and strain the flavoured wine through it into sterilized bottles. Seal the bottles, label, and keep in the refrigerator. The wine is now ready to use.

Makes about 900 ml

WHISKY LIQUEUR WITH HERBS

INGREDIENTS

225 g clear honey

900 ml whisky

2 sprigs of dried fennel

2 sprigs of dried thyme

4 whole cloves

2 cinnamon sticks

1 Put the honey into a saucepan and heat just until warm. Remove from the heat and gradually stir in the whisky until the honey has completely dissolved. Transfer to a sterilized jar and add the herbs and spices. Seal the jar and shake well. Keep in a cool dark place for at least 4 months before using, to allow the flavours to develop. Shake the jar from time to time.

2 Line a funnel with a double layer of muslin and strain the flavoured whisky through it into sterilized bottles. Seal the bottles and label. The whisky is now ready to use.

Makes about 1 litre

CASSIS

INGREDIENTS

450 g blackcurrants

600 ml dry white wine

about 675 g sugar

about 300 ml brandy or gin

1 Lightly crush the blackcurrants in a large glass bowl. Stir in the wine, cover, and leave in a cool dark place for 2 days.

2 Blend the blackcurrant mixture in a food processor or blender. Line a large funnel with a double layer of muslin and strain the liquid through it. Measure the strained liquid. For each 300 ml liquid, weigh 225 g sugar. Put the liquid and sugar into a saucepan, and stir over a low heat until the sugar has completely dissolved. Do not boil. Simmer over a low heat, stirring occasionally, for 45 minutes.

3 Cool, then measure. For 3 parts blackcurrant syrup, add 1 part brandy or gin. Pour into sterilized bottles. Seal the bottles and label. Keep in a cool dark place for 2–3 days before using, to allow the flavours to develop. The cassis is now ready to use.

Makes about 1.15 litres

VARIATIONS Cassis is most commonly made with brandy or gin, but vodka can be used instead. A small cinnamon stick and a whole clove can be added to the syrup to give a touch of spice.

HOT PEPPER SHERRY

INGREDIENTS

6 hot red chillies, fresh or dried

600 ml dry sherry

This is a potent brew, so add a few drops sparingly to stews or sauces.

1 Put the chillies into a sterilized bottle with the sherry. Seal and shake well. Keep in a cool dark place for 2 weeks before using, to allow the flavours to develop. Shake the bottle from time to time. Remove the chillies before using.

Makes about 600 ml

VANILLA BRANDY

INGREDIENTS

2 whole vanilla pods

600 ml brandy

1 Cut the vanilla pods lengthwise to release their flavour. Put them into a sterilized bottle with the brandy. Seal, label, and shake well. Keep in a cool dark place for 3 weeks before using, to allow the flavours to develop. Shake the bottle from time to time.

2 Remove the vanilla pods. The brandy is now ready to use.

Makes about 600 ml

STRAWBERRY SURPRISE
Blushing Strawberry Wine is glamorized with borage flowers.

Fruit Cordials & Syrups

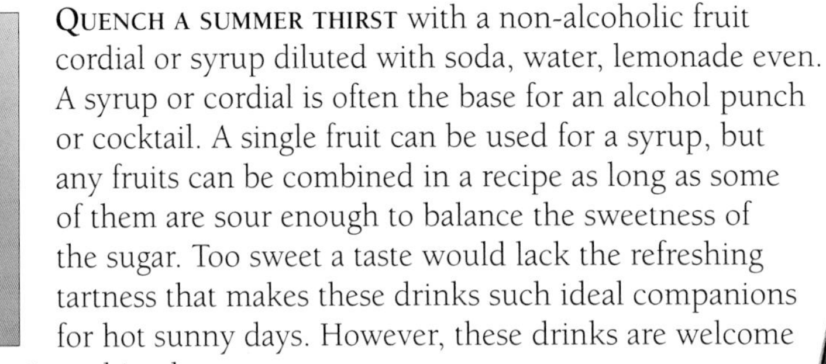

Quench a summer thirst with a non-alcoholic fruit cordial or syrup diluted with soda, water, lemonade even. A syrup or cordial is often the base for an alcohol punch or cocktail. A single fruit can be used for a syrup, but any fruits can be combined in a recipe as long as some of them are sour enough to balance the sweetness of the sugar. Too sweet a taste would lack the refreshing tartness that makes these drinks such ideal companions for hot sunny days. However, these drinks are welcome at any time, thirst knows no season.

The cordials and syrups are concentrated, and a little goes a long way, so they are economical when made in quantity, especially if you are able to pick wild fruits. The best fruits are juicy berries, such as strawberries, raspberries, loganberries, and blackberries. Citrus fruits are good too: utilize both zest and pulp to maximize the citrus flavour. Even tropical fruits make ideal ingredients for cordials and syrups, with kiwi fruit, passion fruit, kumquats, and pineapples giving a subtly fragrant flavour.

Making Fruit Cordials & Syrups

❦ Choose the fruits. Fully ripe or even slightly over-ripe fruits with plenty of flavour are best for cordials and syrups. Granulated sugar is usual choice of sweetener. Make sure that you have enough sugar available before you start, because the exact amount is often calculated after the juice has been extracted from the fruits.

❦ Prepare the fruits; discard any mouldy or damaged parts, and wash and dry them, as necessary. Then extract the juice. This is done by lightly heating, pressing, or squeezing them. The aim is to do this without destroying the flavour, so do not overcook the fruits.

❦ Strain the fruit purée through a double layer of muslin, then lightly squeeze it to extract as much juice as possible.

❦ Add sugar to the juice and dissolve over a very low heat. The mixture is then boiled for about 5 minutes.

❦ Leave the mixture to cool, then pour into sterilized bottles (see page 11), to within 3 mm of the tops.

How to seal and store

Seal tightly with screw-top lids or plastic stoppers. Corks can also be used (see Flavoured Vinegars, page 117). The recipes in this section do not include any special heat treatments nor do they contain sufficient sugar to allow for long storage, so the cordials and syrups should be drunk straightaway, or kept in the refrigerator for only 3–4 weeks. Alternatively, they can be frozen. A convenient way to freeze them is in ice-cube trays: pour in the drink, leaving room for it to expand. When frozen, turn the cubes out into plastic bags and keep in the freezer. One ice cube will be enough for a 240 ml drink, ideal for making in a hurry.

What can go wrong and why

If a cordial or syrup develops a sediment at the bottom of the bottle, it is merely tiny pieces of pulp that have slipped through during straining. Simply pour the cordial or syrup with care so that the sediment is not disturbed.

PINEAPPLE AND LIME SYRUP

This syrup has a taste of the tropics, making it a wonderful base for rum punches and daiquiries. Alternatively, serve it diluted with a little sparkling water as a cooling non-alcoholic drink. Lots of ice cubes and fresh twists of citrus peel will give it added zip.

INGREDIENTS

2 pineapples, total weight about 2 kg

300 ml water

about 175 g sugar

juice of 3 limes

Makes about 600 ml

1 ▶ Slice the tops off the pineapples. Cut off the peel in strips, cutting deep enough to remove the "eyes". Slice the pineapple, and cut out and discard the hard central core from each slice; finely chop the flesh.

2 ▼ Put the pineapple into a saucepan with the water and bring to a boil. Simmer over a low heat for 20 minutes, or until the fruit is soft and pulpy. Mash with a potato masher.

3 ▼ Line a plastic sieve with a double layer of muslin. Strain the pulp through it into a bowl. Gather the corners of the muslin and lightly squeeze to extract all the juice possible. Discard the pulp. Measure the juice. For each 300 ml juice, weigh 100 g sugar.

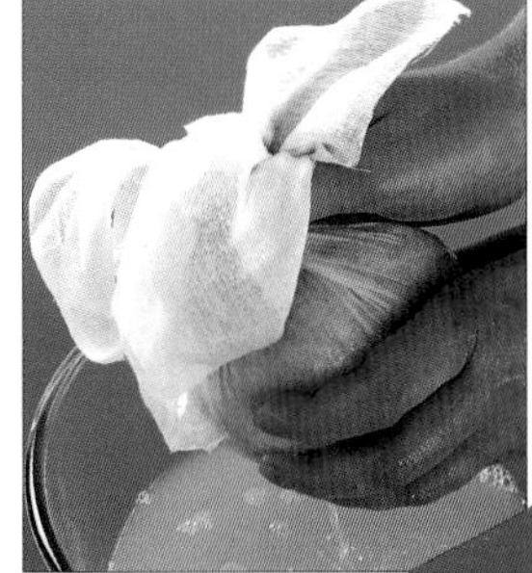

4 ▶ Pour the juice back into the saucepan, add the sugar, and stir over a low heat until the sugar has completely dissolved. Stir in the lime juice.

ZESTY FINISH
Juicy limes add a refreshing tang to pineapple syrup.

5 Bring the mixture to a boil and simmer over a low heat for 5 minutes. Cool for 5 minutes, then pour into sterilized bottles, to within 3 mm of the tops. Seal the bottles, label, and keep in the refrigerator. The syrup is now ready to use.

STRAWBERRY SYRUP

INGREDIENTS

1.4 kg strawberries

sugar

juice of 2 lemons

1 Hull the strawberries and purée them in a food processor. Line a plastic sieve with a double layer of muslin. Strain the pulp through it into a non-metallic bowl. Gather the corners of the muslin and lightly squeeze to extract all the juice possible.

2 Measure the juice. For each 600 ml juice, weigh 350 g sugar. Pour the juice into a saucepan, add the sugar, and stir over a low heat until the sugar has completely dissolved. Stir in the lemon juice. Cool for 5 minutes.

3 Pour into sterilized bottles, to within 3 mm of the tops. Seal the bottles, label, and keep in the refrigerator.

Makes about 1 litre

VARIATIONS
Other strongly acidic fruits, such as raspberries, can be used to make this syrup. Blackcurrants can be used, in which case, add 900 ml cold water while the syrup is cooling in step 2.

LOGANBERRY AND LEMON BALM CORDIAL

INGREDIENTS

1.8 kg loganberries

4 sprigs of lemon balm

600 ml water

sugar

juice of 2 lemons

1 Put the loganberries and the lemon balm into a large saucepan with the water, and bring to a boil. Simmer over a low heat for 8–10 minutes, until the fruit is soft. Mash with a potato masher.

2 Line a plastic sieve with a double layer of muslin. Strain the pulp through it into a non-metallic bowl. Gather the corners of the muslin and lightly squeeze to extract all the juice possible.

3 Measure the juice. For each 600 ml juice, weigh 225 g sugar. Pour the juice back into the saucepan, add the sugar, and stir over a low heat until the sugar has completely dissolved. Stir in the lemon juice. Bring the mixture to a boil and simmer for 5 minutes. Cool for 5 minutes

4 Pour into sterilized bottles, to within 3 mm of the tops. Seal the bottles, label, and keep in the refrigerator.

Makes about 1 litre

KUMQUAT CORDIAL

INGREDIENTS

450 g kumquats

300 ml water

juice of 10 oranges, about 750 ml

sugar

1 Slice the kumquats and put the slices into a saucepan with the water. Bring to a boil and simmer for 30 minutes, until the fruit is very soft.

2 Line a plastic sieve with a double layer of muslin. Strain the pulp through it into a non-metallic bowl. Gather the corners of the muslin and lightly squeeze to extract all the juice possible.

3 Mix the kumquat juice with the orange juice. Measure the total amount of juice. For each 600 ml juice, weigh 225 g sugar. Pour the juice back into the saucepan, add the sugar, and stir over a low heat until the sugar has completely dissolved. Cool for 5 minutes.

4 Pour into sterilized bottles, to within 3 mm of the tops. Seal the bottles, label, and keep in the refrigerator.

Makes about 1 litre

ST. CLEMENT'S CORDIAL

INGREDIENTS

juice of 8 oranges, about 750 ml

juice of 6 lemons, about 350 ml

finely grated zest of 2 oranges and 2 lemons

1 kg sugar

Serve this cordial with an equal amount of iced, still, or sparkling water.

1 Put the juice and grated zest into a saucepan. Add the sugar and stir over a low heat until the sugar has completely dissolved. Increase the heat and bring to just below boiling point. Immediately remove from the heat. Leave to cool completely.

2 Line a plastic sieve with a double layer of muslin. Strain the pulp through it into a non-metallic bowl. Gather the corners of the muslin and lightly squeeze to extract all the juice possible. Pour into sterilized bottles, to within 3 mm of the tops. Seal the bottles, label, and keep in the refrigerator.

Makes about 1.4 litres

BLACKBERRY AND APPLE CORDIAL

INGREDIENTS

1.8 kg blackberries

600 ml apple juice

sugar

juice of 2 lemons

1 Put the blackberries into a large saucepan with the apple juice. Bring to a boil and simmer over a very low heat for 20 minutes, or until the fruit is soft. Mash with a potato masher. Line a plastic sieve with a double layer of muslin. Strain the pulp through it into a non-metallic bowl. Gather the corners of the muslin and lightly squeeze to extract all the juice possible.

2 Measure the juice. For each 600 ml juice, weigh 225 g sugar. Pour the juice back into the saucepan, add the sugar, and stir over a low heat until the sugar has completely dissolved. Stir in the lemon juice. Bring the mixture to a boil and simmer for 5 minutes. Cool for 5 minutes.

3 Pour into sterilized bottles, to within 3 mm of the tops. Seal the bottles, label, and keep in the refrigerator.

Makes about 1 litre

KIWI-PASSIONFRUIT SYRUP

INGREDIENTS

1 kg kiwi fruit

8 passionfruit

sugar

These fragrant tropical fruits are delicious when combined together.

1 Peel and coarsely chop the kiwi fruit. Halve the passionfruit, scoop out the seeds and flesh, and discard the skins. Purée the flesh of both fruits in a food processor or blender.

2 Line a plastic sieve with a double layer of muslin. Strain the pulp through it into a non-metallic bowl. Gather the corners of the muslin and lightly squeeze to extract all the juice possible.

3 Measure the juice. For each 150 ml juice, weigh 225 g sugar. Pour the juice into a saucepan, add the sugar, and stir over a low heat until the sugar has completely dissolved. Bring to a boil and simmer for 5 minutes. Cool for 5 minutes.

4 Pour into sterilized bottles, to within 3 mm of the tops. Seal the bottles, label, and keep in the refrigerator.

Makes about 600 ml

PASSIONFRUIT PULP
Use a small teaspoon to extract the yellow pulp and crunchy black seeds from the passionfruit.

Pickles,
Chutneys,
Relishes &
Mustards

Vegetable & Fruit Pickles

Pickling is the perfect answer to nature's exuberant harvest. Fruits and vegetables can be "put up" in jars and bottles to give a fillip to foods, adding a sharp and spicy accent to ordinary winter fare. The jewel-like colours of ruby beetroots, emerald green beans, and moonstone-shaped garlic cloves make attractive pickles. Fruit pickles can be made from hard or fleshy fruits like apples, peaches, pears, and melons. Probably the most famous vegetable pickles are pickled onions, and cucumber or dill pickles. Pickled onions are a favourite for a pub-type lunch with bread and cheese and a glass of beer or ale. More unusual is pickled garlic, gentled by vinegar, which makes an interesting nibble with drinks.

Pickles have always played an important part in many other cuisines, particularly those of the Middle and Far East. To the pickle-loving Japanese, any meal without pickles is simply incomplete. A favourite of theirs is pickled ginger, which takes little time to prepare, but is infinitely useful as a garnish. Everyone has a list of favourite foods they like to pickle. Oranges, dates, and cherries make exceedingly luxurious pickles, and spiced cranberries enchant with their exquisite pink colour as well as their tart flavour. Green tomato pickles, pickled sweet red capsicums, zucchini flavoured with seeds and spices, or carrots scented with thyme and coriander and sweetened with sugar – the combination of foods for pickling seems almost endless.

Making Vegetable & Fruit Pickles

❦ Select the ingredients. Vinegar, added to prevent spoilage by micro-organisms, is the main ingredient for preserving pickles. Use only best-quality bottled vinegar. Most brands of vinegar have an acetic acid content of 5–7 per cent. Avoid draught vinegar, which may have an acetic acid content of less than 5 per cent. Malt vinegar is the most economical vinegar for pickling. Although its flavour is quite potent, it is the best vinegar for general pickling purposes. Cider and wine vinegar may be recommended where the spicing is more delicate, or when a lighter colour is desired. Spiced Pickling Vinegar (see recipe, page 120) gives a good flavour to pickles. For a more sophisticated taste, vinegars flavoured with herbs or fruits can be used (see Flavoured Vinegars, pages 117–121). The general rule when choosing vinegar, is to match the vinegar to suit the type of vegetable or fruit you are pickling, and to consider the overall colour and flavour you want to create.

❦ Salt, which may be used for salting vegetables before pickling, is another essential preservative for many pickles. Choose sea salt, never table salt which contains chemical additives to keep it free-flowing. Sometimes the vinegar is sweetened; granulated sugar is the usual choice for this, but brown sugars of various kinds may be used for their flavour in certain spiced pickles. Many pickling recipes call for spices: use whole spices because they are easily removed and will also give a clear pickling liquid. Ground spices will cloud the vinegar.

❦ There is no real restriction on the type of vegetables that can be used in pickles, as long as they are firm, even-sized, young, and fresh. If pickling whole items, choose small sizes for easier packing. The most popular pickles are made with beetroots, onions, red and white cabbages, cauliflower, cucumbers, and a mixture of different vegetables. Mushrooms, artichokes, and capsicums also make excellent pickles. Choose fruits that are not over-ripe and are as fresh as possible. Fruits such as melons, peaches, pears, and plums are usually pickled in sweetened vinegar, whereas lemons can be pickled in salt.

❦ Prepare the vegetables or fruits. Using a stainless steel knife, peel and trim the vegetables or fruits as appropriate. Remove any mouldy or damaged parts. If cooking is required, steam the vegetables or cook the fruits in a light syrup.

❦ Some raw vegetables need to be salted or brined before they are pickled. This is to extract the moisture they contain so the vinegar can penetrate the food and preserve it. (Cooked fruits and vegetables do not need to be brined because the cooking process boils off excess water.) The recipe may stipulate dry-brining or wet-brining. For dry-brining, the vegetables are layered with salt in a large non-metallic bowl, left to stand overnight, and stirred occasionally. For wet-brining, a solution of salt and water is poured over the vegetables, which are again left to stand overnight, and stirred occasionally. After brining, the vegetables must be drained, washed, and drained again to remove all the salt, then patted dry with kitchen paper. Avoid using any metal utensils when brining.

❦ Pack the brined or cooked fruits or vegetables into dry sterilized jars (see page 11). Do not pack too tightly, because you need to leave space for the vinegar to surround them. Instead of packing single vegetables separately, layering different vegetables in the same jar can create a decorative effect. Cooked vegetables can be left to stand for an hour in the jars so that any liquid which collects at the bottom can be drained off before proceeding.

❦ Cover the vegetables with vinegar, tapping the jars to prevent air pockets from forming. Fruits are pickled in hot syrup or vinegar or a mixture of both. Leave at least 1.25 cm space at the top of the jar after packing the fruits or vegetables. The vinegar can then be poured in almost to the tops of the jars – ensuring the fruits are well covered, even if some of it evaporates during storage.

❦ Some pickles have a tendency to float to the top of the jar. This can be overcome by putting a piece of crumpled greaseproof paper in the top of the jar above the pickles. Remove after a couple of weeks when the pickles should remain submerged.

How to seal and store

Seal the pickles tightly with non-corrosive, screw-top lids. Pickles should be stored in a cool dark place, such as a cupboard, which is dry and well-ventilated. Vegetable pickles are usually ready to eat after 2–3 weeks, but pickled fruits should be left for about 2 months before using, to allow the flavours to develop. Pickled vegetables and fruits that are covered with pure vinegar will last longer than those which are covered with a combination of water or oil and vinegar. If the vinegar evaporates during storage, top up with more so that the food is always covered.

What can go wrong and why

Yellow spots sometimes appear on pickled onions. This is due to the formation of a harmless substance, and the onions can still be eaten despite their appearance. Certain vegetables, such as zucchini, will have a bitter taste if they are not brined or salted before they are pickled. If pickles do not keep well, it could be because they were not brined for long enough, which means that too much moisture came out of them during storage and diluted the vinegar. Alternatively, the vinegar may not have been up to strength and contained less than 5% acetic acid, which helps to preserve the pickles. Too short a salting or brining time can also make the vinegar cloudy. If the pickles become soft, they may have been left too long before eating. Air pockets will be trapped in the jars if the vegetables or fruits are packed too tightly, and the vinegar has not been able to completely surround them. Gently tapping the jars once they are filled will help remove any pockets that do form. Pickles will go mouldy on top if they are not fully covered with vinegar when being bottled. If the jars are not sealed correctly, vinegar will evaporate, causing the contents of the jars to shrink and dry out. If the lids used are not non-corrosive, then rust will form and it will contaminate the pickles.

DID YOU KNOW? The idea of pickling vegetables and fruits is not a new one. In fact it is one of the most ancient forms of preserving foods, dating back to Greek and Roman times. The Romans had the advantage of having an extensive variety of foods to pickle, which they imported from the lands they conquered in Europe. Onions, lemons, plums, and peaches, as well as herbs, roots, and flowers, would all probably have fallen prey to the large pickling vases used as storage vessels. These were steeped in a mixture of vinegar, oil, and brine, sometimes honey, not unlike the pickles we make today.

EQUIPMENT

Pickles, chutneys, relishes, and mustards use readily available pieces of kitchen equipment, similar to those needed for jams and other preserves. Just a few different utensils are necessary.

Large non-metallic bowls are needed for brining and salting, the initial process in many pickle recipes which helps extract excess water from the vegetables. Everyday china or glass plates can be used to hold the vegetables or fruits down in the brining solution. Kitchen paper is then required to dry the ingredients once they have been rinsed and drained in a plastic colander or sieve. Metal bowls, plates, and sieves should be avoided, particularly if they are to come into contact with vinegar, because vinegar will corrode them and give the pickle a disagreeable taste. Small stainless steel pans are ideal for heating vinegars and vinegar syrups. Chutneys and relishes are best made in a good-quality stainless steel preserving pan. Make sure it is spotless and that the pan is neither pitted nor buckled. Avoid brass or copper preserving pans, because the vinegar will corrode them.

If whole spices are to be added to the mixture, but need to be removed later, a piece of muslin is most useful for tying them into a bag. Use a double thickness of muslin and secure to the pan handle with a long piece of string, so that the bag can be lifted out easily.

Greaseproof paper is often used in pickles to keep the ingredients under the level of the vinegar, and this should be removed after 1–2 weeks storage. Although ingredients, such as sugar and salt, used in making pickles, chutneys, relishes, and mustards, are to some degree natural preservatives, using sterilized containers and lids for storage will also protect against spoilage. Make sure that glass jars are as clean as possible, not cracked, and can be made airtight with non-corrosive, screw-top lids.

PICKLED ONIONS

INGREDIENTS

1–1.15 kg pickling onions, depending on size

100 g sea salt

1.15 litres water

750 ml strained Spiced Pickling Vinegar *(see recipe, page 120)*

In this recipe, special pickling onions have been used, but you can use shallots instead if you prefer.

1 Trim and peel the onions, leaving a little of the fibrous root attached so that they remain whole and do not fall apart. Put them into a large non-metallic bowl.

2 Dissolve the salt in the water and pour over the onions. Cover and leave to stand for 24 hours, stirring from time to time.

3 Rinse the onions in a plastic sieve or colander under cold running water. Drain well and dry on kitchen paper.

4 Pack the onions into sterilized jars, to within 2.5 cm of the tops. Pour in the vinegar, to cover them by 1.25 cm. Gently tap the jars to remove any air pockets.

5 Seal the jars and label. Keep in a cool dark place for 3 weeks before using, to allow the flavours to develop.

Fills about three 300 ml jars

COOK'S TIP If the vegetables in any of these recipes show a tendency to float, put crumpled greaseproof paper in the top of the jar to hold the vegetables under the liquid. Remove the paper after 1 week.

PICKLED HORSERADISH

INGREDIENTS

175 g fresh horseradish

350 ml strained Spiced Pickling Vinegar *(see recipe, page 120)*

75 g sugar

1 tsp sea salt

1 Peel and finely grate the horseradish, then check the weight. You should have 100 g grated horseradish.

2 Put the vinegar, sugar, and salt into a saucepan and bring to a boil. Stir over a low heat, with a wooden spoon, until the sugar has completely dissolved, then boil for 2 minutes. Add the horseradish and simmer for 1 minute.

3 Pack the horseradish and vinegar mixture into warmed sterilized jars, to within 3 mm of the tops.

4 Seal the jars and label. Keep in a cool dark place for 1 week before using, to allow the flavours to develop.

Fills about two 175 ml jars

MOROCCAN PICKLED CAPSICUMS

INGREDIENTS

1 kg red and green capsicums

50 g sea salt

2 small heads of garlic

300 ml olive oil

300 ml white wine vinegar

1 Cut around the core of each capsicum and pull it out. Quarter each capsicum, scrape out the seeds, and cut away the white ribs. Sprinkle the inside of each quarter liberally with the salt.

2 Peel the garlic cloves. Mix the olive oil and wine vinegar together. Pack the capsicums and garlic, in layers, into a sterilized jar, to within 2.5 cm of the top.

3 Pour in the oil and vinegar mixture, to cover the capsicums and garlic by 1.25 cm.

4 Gently tap the jar to remove any air pockets. Seal the jar and label. Keep in the refrigerator and use within 1–2 weeks.

Fills one 1.25 litre jar

MIXED VEGETABLE PICKLES

A variety of crisp vegetables look attractive when neatly layered in a tall jar and covered with Spiced Pickling Vinegar. Use them to create a colourful winter salad, or as an accompaniment to creamy dips.

Make sure the vegetables are well covered with vinegar

INGREDIENTS

450 g green beans

450 g small pickling onions

1 small cauliflower, weighing about 350 g

350 g zucchini

225 g peeled carrots

100 g sea salt

450 ml strained Spiced Pickling Vinegar (see recipe, page 120)

Fills one 1.15 litre jar and one 600 ml jar

1 ▲ Trim the beans and cut them into neat 2.5 cm lengths. Trim and peel the onions, leaving them whole, but with a little of the fibrous root attached so that they do not fall apart. Cut the florets from the cauliflower and break into 1.25 cm pieces. Trim the zucchini and carrots and cut them into 3 mm thick slices.

2 ▸ Layer the vegetables with the salt in a large non-metallic bowl. Put a plate on top to keep the vegetables lightly pressed down, and leave for 24–48 hours, to draw out the excess moisture.

3 Rinse the vegetables in a plastic sieve or colander under cold running water until all the salt is washed off. Drain the vegetables well and dry on kitchen paper.

4 Neatly pack the vegetables in separate layers into sterilized jars, to within 2.5 cm of the tops.

5 ▲ Slowly pour in the pickling vinegar, to cover the mixed vegetables by 1.25 cm. Gently tap the jars to remove any air pockets. Seal the jars and label. Keep in a cool dark place for 6 weeks before using, to allow the flavours to develop.

Pickled Baby Vegetables

INGREDIENTS

1.15 kg mixed baby vegetables

100 g sea salt

800 ml strained Spiced Pickling Vinegar (see recipe, page 120)

Choose baby vegetables which will maintain their crispness during pickling. Carrots, sweetcorn, and snow peas are possible choices.

1 Trim all the vegetables and peel, if necessary. Layer the vegetables with the salt in a large non-metallic bowl. Put a plate on top to keep them lightly pressed down, and leave to stand overnight, to draw out the excess moisture.

2 Rinse the vegetables in a plastic sieve or colander under cold running water. Drain well and dry on kitchen paper.

3 Pack the vegetables into sterilized jars, to within 2.5 cm of the tops. Slowly pour in the pickling vinegar, to cover the baby vegetables by 1.25 cm. Gently tap the jars to remove air pockets.

4 Seal the jars and label. Keep in a cool dark place for 4 weeks before using, to allow the flavours to develop.

Fills about two 900 ml jars

PICKLED JERUSALEM ARTICHOKES Do not salt and leave to stand overnight, but simmer in 600 ml salted water until tender. Drain, cool, and dry, then continue from step 3.

Dill Pickles with Garlic

INGREDIENTS

3 large garlic cloves

10–12 pickling cucumbers

1 bunch of fresh dill about 7 g

600 ml water

900 ml white wine vinegar

75 g sea salt

2 tsp Pickling Spice (see recipe, page 126)

1 tsp dill seeds

1 tsp black peppercorns

There are many recipes for dill pickles, which are also known as kosher pickles. In this recipe water is mixed with vinegar to make a mild pickle. Other flavourings can also be used, such as onions, fennel, or cloves.

1 Peel the garlic cloves and pack them into a sterilized jar with the cucumbers and sprigs of dill.

2 Pour the water and vinegar into a saucepan and add the salt, pickling spice, dill seeds, and peppercorns. Bring to a boil, and boil rapidly for 3 minutes. Leave to cool.

3 Pour the cooled mixture over the cucumbers, to cover them by 1.25 cm. If there is too much liquid, add any remaining spices to the jar and discard the excess liquid.

4 Seal the jar and label. Keep in the refrigerator for 3 weeks before using, to allow the flavours to develop.

5 After a day or two, check that the cucumbers are still well covered. If they are protruding out of the vinegar, push them down into the liquid and hold them down with a crumpled piece of greaseproof paper. Remove the paper after 2 weeks.

Fills one 2 litre jar

VARIATION Instead of leaving the cucumbers whole, thinly slice them and layer them in sterilized jars, with 2 peeled and sliced onions, 2 peeled garlic cloves, and the dill. Continue as for whole cucumbers.

Pickled Cauliflower with Capsicums

Ingredients

2 cauliflowers

1 red capsicum

150 g sea salt

900 ml water

1 cinnamon stick, broken into 3 pieces

3 blades of mace

900 ml strained Spiced Pickling Vinegar (see recipe, page 120)

1 Cut the florets from the cauliflowers and discard the stalks. Put the florets into a large non-metallic bowl. Core, seed, and dice the capsicum and add to the bowl with the cauliflower.

2 Dissolve the salt in the water and pour over the vegetables. Weight down with a plate to keep the vegetables submerged. Leave to stand for 24 hours.

3 Rinse the cauliflower and capsicum in a plastic sieve or colander under cold running water. Drain well and dry on kitchen paper.

4 Pack the vegetables into sterilized jars, to within 2.5 cm of the tops, adding a piece of cinnamon stick and a blade of mace to each jar. Pour in the pickling vinegar, to cover the vegetables by 1.25 cm.

5 Seal the jars and label. Keep in a cool dark place for 4 weeks before using, to allow the flavours to develop.

Fills about three 300 ml jars

Zucchini Pickles

Ingredients

1.4 kg zucchini

450 g onions

225 g sea salt

2.5 litres water

450 ml white vinegar

225 g sugar

2 tsp mustard seeds

1 tsp celery seeds

1 tsp whole allspice

1 Trim the zucchini and peel and trim the onions. Slice both vegetables thinly and place them in a large non-metallic bowl.

2 Dissolve the salt in the water and pour over the vegetables. Leave to stand for 3 hours, stirring from time to time.

3 Mix all the remaining ingredients in a saucepan. Stir over a low heat, with a wooden spoon, until the sugar has completely dissolved.

4 Rinse the zucchini and onions in a plastic sieve or colander under cold running water. Drain well and dry on kitchen paper. Put into a preserving pan. Pour the hot vinegar over the vegetables and leave to stand, off the heat, for 1 hour.

5 Put the pan over a medium heat, and bring to a boil. Boil for 3 minutes, then remove from the heat.

6 Transfer the vegetables to warmed sterilized jars, to within 2.5 cm of the tops. Pour in the vinegar, to cover by 1.25 cm. Seal the jars and label. Keep in a cool dark place for 2 weeks before using, to allow the flavours to develop.

Fills about two 750 ml jars

SWEET CAPSICUM PICKLES

Capsicums offer a riot of colour when used in making pickles. Extra flavour comes from slivers of sun-dried tomatoes, and spiced vinegar which has a hint of sweetness. Serve these pickles with cold meats or on an antipasto platter.

INGREDIENTS

4 large capsicums (red, green, orange, and yellow)

1 onion

175 g sea salt

350 ml strained Spiced Pickling Vinegar (see recipe, page 120)

50 g sugar

25 g sun-dried tomatoes

Fills one 1.15 litre jar

1 ◀ Cut around the core of each capsicum and pull it out. Halve each capsicum lengthwise and scrape out the seeds. Cut away the white ribs on the inside. Cut each half lengthwise into 3 mm wide strips. Peel and finely chop the onion.

2 Layer the capsicums, onion, and salt in a large non-metallic bowl. Put a plate on top to keep the vegetables lightly pressed down, and leave for 24 hours, covered, to draw out the excess moisture. Toss from time to time.

3 Rinse the vegetables in a plastic sieve or colander under cold running water, until all the salt is washed off. Drain and dry on kitchen paper.

4 Gently heat the pickling vinegar and sugar in a small saucepan, stirring until the sugar has completely dissolved. Cool.

5 Put the sun-dried tomatoes into a bowl and cover with boiling water. Leave for 5 minutes, then drain well.

6 ▲ Cut the tomatoes into thin shreds. Mix with the other vegetables and pack them into a sterilized jar, to within 2.5 cm of the top. Pour in the vinegar, to cover the vegetables by 1.25 cm. Gently tap the jar to remove any air pockets.

7 Seal the jar and label. Keep in a cool dark place for 6 weeks before using, to allow the flavours to develop.

Sweet and Sour Carrots

Ingredients

1 kg small carrots

100 g sea salt

2 garlic cloves

300 ml white wine vinegar

100 g brown sugar

2 tsp ground cinnamon

1 tsp ground coriander

1 tsp dried thyme

1 small dried red chilli

These oriental-style carrots can be served on a platter of mixed hors d'oeuvre. Olives, artichoke hearts, and marinated capsicums are excellent accompaniments.

1 Peel and thinly slice the carrots. Layer them with the salt in a large non-metallic bowl. Put a plate on top to keep the carrots lightly pressed down, and leave for 24 hours, covered, to draw out the excess moisture. Toss from time to time.

2 Peel the garlic cloves and put them into a saucepan with all the remaining ingredients. Stir over a low heat until the sugar has completely dissolved. Remove from the heat and leave the mixture to stand until required.

3 Rinse the carrots in a plastic sieve or colander under cold running water, until all the salt is washed off. Drain and dry on kitchen paper.

4 Pack the carrots into sterilized jars, to within 2.5 cm of the tops. Slowly pour in the vinegar, to cover the carrots by 1.25 cm. Gently tap the jars to remove any air pockets.

5 Seal the jars and label. Keep in a cool dark place for 1 week before using, to allow the flavours to develop.

Fills about two 350 ml jars

Pickled Red Cabbage

Ingredients

1 small red cabbage, weighing about 1.15 kg

75 g sea salt

1 tsp sugar

300 ml strained Spiced Pickling Vinegar (see recipe, page 120)

1 Remove any discoloured outer leaves from the cabbage. Cut the cabbage into quarters, and discard the tough central core. Shred the cabbage finely.

2 Layer the cabbage with the salt in a large non-metallic bowl. Put a plate on top to keep the cabbage lightly pressed down, and leave for 24 hours, covered, to draw out the excess moisture.

3 Rinse the cabbage in a plastic sieve or colander under cold running water, until all the salt is washed off. Drain and dry on kitchen paper.

4 Pack the cabbage into sterilized jars, sprinkling with a little sugar as you go, to within 2.5 cm of the tops. Slowly pour in the pickling vinegar, to cover the cabbage by 1.25 cm. Gently tap the jars to remove any air pockets.

5 Seal the jars and label. Keep in a cool dark place for 2–3 weeks before using, to allow the flavours to develop.

Fills about two 600 ml jars

Variation
A quick version of Pickled Red Cabbage can be made if the red cabbage is not salted. However, it will not remain as crisp, and must be eaten within 2–3 months.

PICKLED DICED BEETROOT

INGREDIENTS

1 kg uncooked beetroots

1 tsp sea salt

1 tsp freshly grated horseradish (optional)

1.15 litres white wine vinegar

This recipe is for cubes of beetroot. If slices are preferred, follow the recipe as directed and, when the beetroots are cool, peel and slice them. Pack the slices into sterilized jars and fill with vinegar.

1 Wash the beetroots, taking care not to damage their skins.

2 Put the beetroots into a large saucepan and cover with water. Add the salt. Simmer, covered, for 1½–2 hours, or until the beetroots are tender. Do not pierce them to test or the beetroots will "bleed" and lose their colour. Leave to cool.

3 Peel the beetroots, and cut into small cubes. Pack the diced beetroot into sterilized jars, with the horseradish if using, to within 2.5 cm of the tops.

4 Put the vinegar into a pan and bring to a boil. Pour the boiling vinegar over the the beetroot, to cover it by 1.25 cm. Gently tap the jars to remove any air pockets.

5 Seal the jars and label. Keep in a cool dark place for 3 weeks before using, to allow the flavours to develop.

Fills about three 300 ml jars

TANGY PICKLED MUSHROOMS

INGREDIENTS

675 g mushrooms

7.5 cm piece of fresh root ginger

1 lemon

1 onion

1 litre white wine vinegar

3 tsp sea salt

1 tsp black peppercorns

Although mushrooms make a great pickle, they do not keep as long as many other vegetables. Make small amounts, keeping them in the refrigerator, and eat them quickly.

1 Wipe the mushrooms with a damp cloth and trim the stalks so that they are even with the caps. Do not wash or peel them. Put the mushrooms into a saucepan.

2 Peel the root ginger with a small sharp knife, and cut it into quarters. Pare the zest from the lemon, with a vegetable peeler, and cut the zest into thin strips. Add the ginger and lemon strips to the mushrooms.

3 Peel and thinly slice the onion and add to the pan with all the remaining ingredients. Bring the mixture to a boil, and simmer for 15–20 minutes, or until the mushrooms are tender.

4 Lift the mushrooms out of the cooking liquid with a slotted spoon and pack into sterilized jars.

5 Strain the liquid, pour it back into the pan, and return to a boil. Pour it over the mushrooms, to cover them by 1.25 cm.

6 Seal the jars, label, and keep in the refrigerator. The mushrooms are now ready to use.

Fills about three 300 ml jars

COOK'S TIP Use close-capped or tiny button mushrooms for this recipe. They should be completely fresh and pure white, with pale pink gills.

SWEET PICKLED BEANS

INGREDIENTS

675 g green beans

3 tsp Pickling Spice *(see recipe, page 126)*

3 tsp black peppercorns

3 tsp white wine vinegar

75 g sugar

1 bay leaf

1 garlic clove

1 onion

1 red capsicum

4 large sprigs of dill

Serve pickled beans as an hors d'oeuvre or with egg mayonnaise.

1 Top and tail the beans and cut into 7.5 cm lengths. Put them into a pan of boiling water and boil for 1 minute. Drain, rinse under cold running water, and drain again.

2 Put the pickling spice and peppercorns on a square of muslin and tie up tightly into a bag with a long piece of string. Tie the bag to the handle of a saucepan. Put the vinegar, sugar, bay leaf, and garlic clove into the saucepan.

3 Stir the mixture over a low heat, with a wooden spoon, until the sugar has completely dissolved. Bring to a boil and simmer for 10 minutes. Remove and discard the muslin bag, bay leaf, and garlic clove. Leave to cool.

4 Meanwhile, peel and dice the onion. Core, seed, and dice the red capsicum. Mix the two together and set aside.

5 Pack the beans upright into sterilized jars. Spoon in the onion and capsicum mixture to within 2.5 cm of the tops. Add 2 sprigs of dill to each jar and pour in the sweetened vinegar to cover the beans by 1.25 cm.

6 Seal the jars and label. Keep in a cool dark place for 2 weeks before using, to allow the flavours to develop.

Fills about two 600 ml jars

JAPANESE PICKLED GINGER

INGREDIENTS

225 g fresh root ginger

sea salt

240 ml rice vinegar

3 tsp sugar

a few drops red food colouring

In Japan, pickled ginger is always served with sushi *– vinegared rice garnished with raw fish and shellfish. It also goes well with other seafoods and poultry. Traditionally, fresh plum juice is used to subtly colour the pickling liquid, but red food colouring works just as well.*

1 Peel the ginger and cut it along the grain into the thinnest possible slivers. Put them into a bowl, cover with cold water, and leave to stand for 30 minutes.

2 Drain the ginger, and put it into a saucepan of boiling water. Bring back to a boil, drain again, and let cool. Return the ginger to the bowl and sprinkle lightly with salt.

3 In a saucepan, combine the vinegar and sugar. Stir over a low heat until the sugar has completely dissolved. Stir in a few drops of red food colouring. Pour the vinegar mixture over the ginger, making sure it is completely submerged. Cover the bowl and let the ginger stand in a cool dark place for 2 weeks before using, to allow the flavours to develop.

4 Transfer the ginger and liquid to a sterilized jar. Seal the jar, label, and keep in the refrigerator.

Fills one 350 ml jar

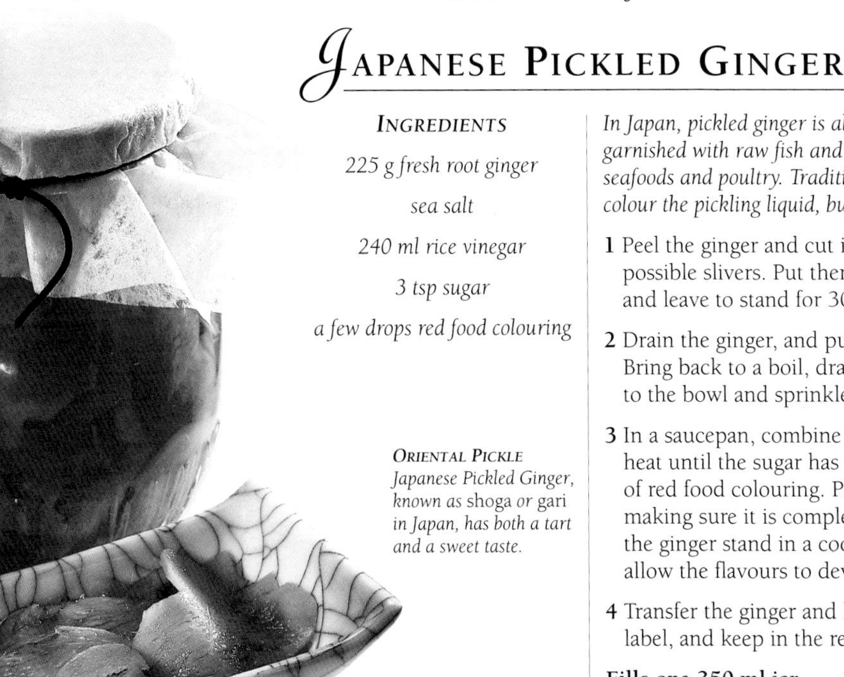

ORIENTAL PICKLE
Japanese Pickled Ginger, known as shoga *or* gari *in Japan, has both a tart and a sweet taste.*

FIVE-SPICE PEACHES

Bottle up the taste of summer with these juicy peaches. Spiced and enveloped in a fragrant sweet and sour syrup, they make an ideal partner for cold meats, especially pork and ham. Pack them into one large jar, or two smaller ones, for individual gifts.

INGREDIENTS

1.8 kg peaches
600 ml white wine vinegar
12 black peppercorns
1 tsp whole cloves
4 cardamom pods
2 cinnamon sticks
2 star anise
1 kg sugar
Fills one 1.15 litre jar

1 ◄Bring a saucepan of water to a boil and put each peach into it for 30–40 seconds. The timing will depend on the ripeness of the fruit. Remove the peaches with a slotted spoon and transfer to a bowl of iced water. Leave to cool, then drain and dry well.

PICKING PEACHES
Select your peaches for pickling at the height of the season when their skins blush with a rosy bloom.

2 ▲Halve the peaches, using the indentation on one side as a cutting guide. Stone and peel them. Discard the stones and skin and set the peaches aside.

3 ▶Bring the wine vinegar and spices to a boil in a preserving pan. Add the sugar and stir, over a low heat, until it has completely dissolved.

4 ▼Boil the mixture for 2 minutes. Add the peach halves and simmer for 4–5 minutes, or until the peaches are tender when pierced with a knife.

SPLENDID SPICES
Whole spices are used rather than ground ones to give the peaches a crystal-clear syrup in which to soak and mature.

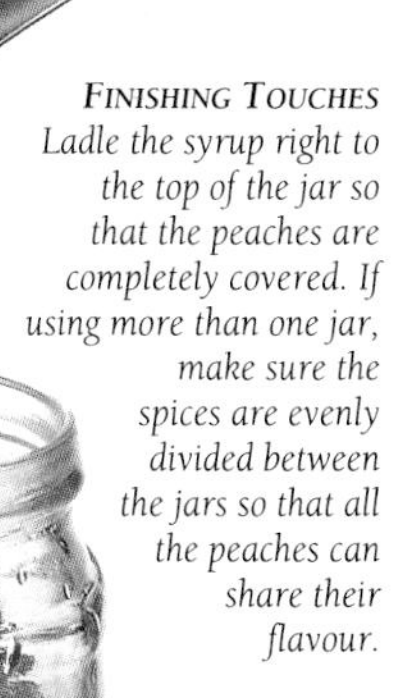

FINISHING TOUCHES
Ladle the syrup right to the top of the jar so that the peaches are completely covered. If using more than one jar, make sure the spices are evenly divided between the jars so that all the peaches can share their flavour.

5 ▲Transfer the peaches to a sterilized jar, to within 1.25 cm of the top. Boil the syrup for a further 2–3 minutes, until it has slightly reduced.

6 ▶Pour syrup over the peaches. Seal the jar and label. Keep in a cool dark place for 2 months before using, to allow the flavours to develop.

PICKLED CHERRIES

INGREDIENTS

1.75 litres white wine vinegar

3 tbsp sea salt

1 kg cherries

4 sprigs of fresh tarragon

12 black peppercorns

1 Pour half of the vinegar into a saucepan and add half of the salt. Boil for 3 minutes; remove from heat and leave to cool.

2 Prick the cherries all over with a sterilized needle or a wooden cocktail stick. Pack into 2 sterilized jars, with 2 sprigs of tarragon and 6 peppercorns in each jar, to within 2.5 cm of the tops.

3 Pour in the cold salted vinegar to cover the cherries by 1.25 cm, making sure that there are no air pockets. Add crumpled greaseproof paper to the tops of the jars, to keep the cherries under the surface of the vinegar. Seal the jars. Keep in a cool dark place for 1 week.

4 Repeat step 1 with the remaining vinegar and salt. Remove the greaseproof paper and drain off the old salted vinegar from the cherries. Discard the tarragon sprigs. Pour in the fresh salted vinegar as in step 3.

5 Seal again and label. Keep in a cool dark place for 3 weeks before using, to allow the flavours to develop.

Fills about two 900 ml jars

SPICED APPLES WITH ROSEMARY

INGREDIENTS

450 ml cider vinegar

120 ml honey

12 large crisp eating apples (Granny Smith)

6 sprigs of fresh rosemary

1 tsp allspice berries

This pickle is delicious served with pork, poultry, or ham, as an alternative to apple sauce.

1 Pour the vinegar into a saucepan and add the honey. Stir over a low heat until the honey has completely dissolved. Bring to a boil, and simmer for 2–3 minutes.

2 Peel, halve, and core the apples. Cut the apple flesh into quarters and then into eighths. Add to the pan of vinegar and honey. Simmer for 8–10 minutes, or until the apples are tender but not too soft.

3 Carefully pack the apples into 3 warmed sterilized jars, to within 2.5 cm of the tops, adding 2 sprigs of rosemary and a few allspice berries to each jar. Pour in the vinegar mixture, to cover the apples by 1.25 cm, making sure that there are no air pockets. Seal the jars and label.

4 Keep in a cool dark place for 5–6 days before using, to allow the flavours to develop.

Fills about three 600 ml jars

DID YOU KNOW? Honey ranges in appearance from light and clear, to thick and opaque; the flavour depends on the flower from which the pollen came. Any commercial honey can be used in this recipe.

Spiced Oranges

Ingredients

4 large seedless oranges

450 g sugar

300 ml white wine vinegar

1/2 tsp whole cloves

1 cinnamon stick

3 blades of mace

1 With a sharp chef's knife, cut the unpeeled oranges into slices about 5 mm thick.

2 Put the orange slices into a saucepan and pour in just enough water to cover. Half cover the pan with a lid and simmer for about 1 hour, or until the orange slices are soft. Remove the slices carefully and drain well in a plastic sieve or colander.

3 Combine the sugar and vinegar in the saucepan and stir over a low heat until the sugar has completely dissolved. Add the spices to the syrup and boil for 5 minutes. (If you wish to remove the spices later, put them into a muslin bag before adding to the sugar and vinegar.)

4 Add the orange slices to the syrup. Simmer, covered, for 15–20 minutes, or until the orange slices are semi-transparent and the syrup has reduced and thickened.

5 Remove the orange slices from the syrup, using a slotted spoon, and arrange neatly in warmed sterilized jars, to within 1.25 cm of the tops. Discard the muslin bag, if used. Pour the syrup and spices over the orange slices, making sure that there are no air pockets.

6 Seal the jars and label. Keep in a cool dark place for 6 weeks before using, to allow the flavours to develop.

Fills about two 600 ml jars

Pickled Dates

Ingredients

1 kg fresh dates

3 tsp sea salt

600 ml white wine vinegar

12 black peppercorns

12 whole cloves

2 cinnamon sticks

Serve this pickle with a Cheddar cheese ploughman's lunch.

1 Put the dates into a pan of boiling water. Immediately remove from the heat and leave to stand for 2 minutes. Drain and peel. Cut the dates in half lengthwise; remove and discard the stones.

2 Layer the dates with the salt in 2 sterilized jars, to within 2.5 cm of the tops.

3 Put all the remaining ingredients into a saucepan and bring to a boil. Boil for 1 minute, then pour the mixture over the dates to cover them by 1.25 cm, making sure that there are no air pockets between the dates.

4 Seal the jars and label. Keep in a cool dark place for 2 weeks before using, to allow the flavours to develop.

Fills about two 600 ml jars

Variation Adapt the above recipe to make Pickled Walnuts. First cover the nuts with a standard brine: use 1 1/2 tbsp salt and 600 ml water for each 450 g shelled walnuts. Leave for 3 days. Drain, and cover the walnuts with fresh brine. Leave for 1 week. Drain the walnuts and pack into sterilized jars. Boil the vinegar and spices for 1 minute, cool, and pour into the jars to cover the walnuts. Finish as directed, keeping for 5–6 weeks before using.

PEPPERED KIWI FRUIT

Chunks of tropical kiwi fruit blend with apple in this exotic pickle. Very light cooking in a mildly peppered syrup ensures the texture and flavour of the fruits are preserved. Try the pickle as an unusual cooling accompaniment to hot curries.

INGREDIENTS

8 kiwi fruit, total weight about 675 g

1 tsp green peppercorns, in brine

1 tsp mustard seeds

600 ml clear unsweetened apple juice

150 ml white wine vinegar

225 g sugar

2 large crisp eating apples (Granny Smith)

Fills about three 300 ml jars

1 ▶Peel the kiwi fruit. Cut each fruit lengthwise into quarters, then cut each quarter crosswise into thirds. Rinse the peppercorns in a small sieve. Drain and pat dry. Lightly crush the peppercorns and mustard seeds. Set them aside.

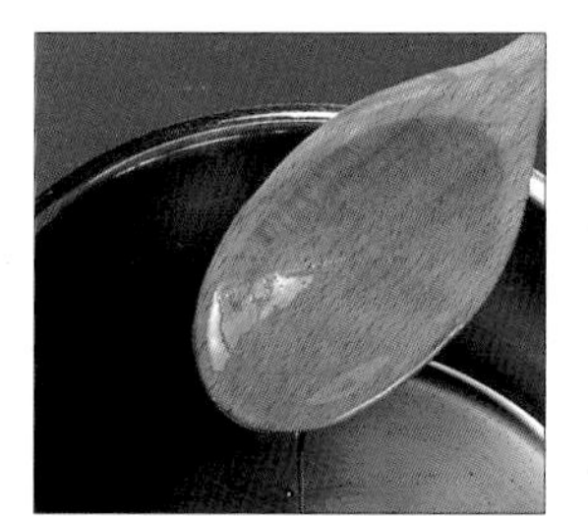

2 ◀Pour the apple juice and wine vinegar into a saucepan and bring to a boil. Add the sugar and stir over a low heat, with a wooden spoon, until it has completely dissolved. Simmer, without stirring, for 10–15 minutes, or until the syrup has slightly reduced.

3 Meanwhile, peel and quarter the apples, and discard the cores. Cut the flesh into piece the same size as the kiwi fruit.

4 Put the apple pieces, crushed peppercorns, and mustard seeds into the saucepan with the syrup. Stir gently to mix, taking care not to break up the apple.

5 ▲Bring the mixture to a boil, and simmer, stirring occasionally, for 2 minutes. Add the kiwi fruit and simmer for a further 2 minutes, or until the fruits are just tender when pierced with a sharp knife.

6 Immediately spoon the fruits and syrup into warmed sterilized jars, to within 3 mm of the tops. Seal the jars and label. Keep in a cool dark place for 1 week before using, to allow the flavours to develop.

SPICED PICKLED PEARS

INGREDIENTS

1.4 kg pears

2 cinnamon sticks

1 tsp allspice berries

1 tsp whole cloves

2 blades of mace

2 small pieces of peeled dried root ginger

225 g sugar

480 ml cider vinegar

These aromatically spiced pears are excellent for serving with duckling, or any roast or cold meats, poultry, or game. They are also good chopped and added to curries at the last minute.

1 Peel the pears. Quarter them if large, or halve them if small. Core the pears and put them into a saucepan with just enough water to cover. Bring to a boil, and simmer for 5 minutes.

2 Drain the pears in a plastic sieve set over a large measuring jug. Measure the liquid and add enough water to make 480 ml. Pour the liquid back into the saucepan and stir in all the remaining ingredients. Simmer for 5 minutes.

3 Add the pears to the pan and continue to simmer for 30 minutes, or until they are translucent.

4 Transfer the pears to a warmed sterilized jar, to within 1.25 cm of the top. Pour the syrup over the pears to cover. If using more than one jar, make sure that the spices are evenly divided between the jars.

5 Seal the jar and label. Keep in a cool dark place for at least 1 month before using, to allow the flavours to develop.

Fills one 1.15 litre jar

VARIATION To make Spiced Pickled Apricots, first halve and stone the apricots, then proceed as for the pears. Reduce the final cooking time, cooking the apricots just until they are soft.

SPICED PRUNES WITH EARL GREY TEA

INGREDIENTS

1 kg prunes

900 ml cold Earl Grey tea

300 ml white wine vinegar

1 cinnamon stick

225 g sugar

16 allspice berries

6 whole cloves

2 blades of mace (optional)

Spiced prunes are delicious with roast pork or hare. Their aromatic liquor can also be added sparingly to sauces and casseroles.

1 Put the prunes into a large non-metallic bowl, pour in the tea, cover, and leave to soak overnight.

2 Put the prunes and liquid into a saucepan. Bring to a boil, and simmer for 15–20 minutes, or until the prunes are tender. Transfer the prunes to warmed sterilized jars, to within 1.25 cm of the tops.

3 Add the wine vinegar to the saucepan. Break the cinnamon stick in half and add it to the saucepan with the remaining ingredients. Stir over a low heat until the sugar has completely dissolved. Increase the heat and boil the syrup for 2–3 minutes, or until it has reduced and thickened.

4 Pour the syrup over the prunes to cover. Make sure the spices are evenly divided between the jars.

5 Seal the jars and label. Keep in a cool dark place for at least 2 weeks before using, to allow the flavours to develop.

Fills about two 750 ml jars

DID YOU KNOW? Earl Grey tea is a scented tea, named after the 2nd Earl Grey, for whom it was created. Earl Grey tea is a blend of China and Darjeeling teas, and is delicately flavoured with oil of bergamot.

CHUTNEYS

ORIGINALLY FROM INDIA, where their Hindu name, *chatni,* means "strongly spiced", chutneys are much more than that – rich concoctions of fruits, vegetables, herbs, and spices, that are slow-cooked with vinegar and sugar to create a sweet-sour condiment. The most popular spices are the aromatic ones – cinnamon, ginger, allspice, and cardamom. Mango chutney is surely the greatest favourite, and is a good way to preserve the perfumed flavour of mangoes with their all-too-short season.

Chutneys can be freshly made to eat right away, but the cooked version is the most commonly known. Cooked chutneys are an ideal side dish, and they go with almost anything – hot or cold meats and poultry, cheeses, vegetables, and eggs. They keep well, improving with age.

A chutney invites improvization. Although mango chutney is a classic, magnificent chutneys can be made with nectarines, peaches, pears, apricots, beetroots, green and red tomatoes, pineapples, pawpaws – the list is almost endless. They can be very hot, if made with chillies, or mild, with their sweetness predominating. The perfect time for chutney making is when there is a glut of fruits and vegetables on the market, and there is no reason why a gifted cook should not experiment with new combinations of ingredients – dates with oranges, green tomatoes with apples, and marrow with apricots are but a few of the possibilities.

MAKING CHUTNEYS

❦ Select the ingredients. Chutney is perfect for using up misshapen and bruised fruits and vegetables that are not suitable for other preserves in which appearance is important. The produce can also be riper than that used for jam making, but not over-ripe. Other ingredients added for flavour are spices, vinegar, dried fruits, nuts, and sugar. White sugar, or brown (for its colour and flavour) can be used, so too can honey, golden syrup, and molasses, but in small quantities only, as they can crystallize during storage.

❦ Wash the fruits and vegetables and, if necessary, peel or skin. Chop the fruits and vegetables by hand, or in a food processor for a finer, less chunky chutney.

❦ Long slow cooking is best for chutneys, so that the ingredients can break down and all the flavour be extracted. It may be necessary to simmer hard- or thick-skinned fruits or vegetables, such as unripe pears, carrots, or gooseberries, in vinegar first to soften them.

❦ The chutney is ready when it has reduced and thickened to a jam-like consistency. There should be no runny vinegar visible when a spoon is drawn in a line across the bottom of the pan. Spoon the chutney into warmed sterilized jars (see page 11), to within 3 mm of the tops, and stir with a metal skewer to remove any air pockets.

How to seal and store

Seal the jars very tightly with non-corrosive lids. Metal screw-top lids with plastic linings are the best choice, because they ensure that the vinegar does not evaporate and dry out the chutney. Label the jars. Although chutneys can be eaten immediately, most are best left for 1 month before eating, to allow the flavours to develop. Chutneys generally keep for up to 1 year if stored in a cool dry and dark place, but some chutneys have different storage times, which are given in individual recipes.

What can go wrong and why

If the chutney dries out or shrinks in the jar, it has been over-boiled, not covered tightly enough, or stored in a warm place. Mouldy chutney is caused by insuffiicent vinegar, under-cooking, or the use of unsterilized jars. Liquid on the surface of a chutney is a result of insufficient boiling down of the mixture.

DARK MANGO CHUTNEY

Some mango chutneys are sweet and light – this one is mysteriously dark and spicy. No individual flavour dominates, the ingredients just blend together in a simply delightful way.

INGREDIENTS

1.4 kg mangoes

1½ tbsp salt

100 g tamarind pulp

75 g fresh root ginger

15 g dried red chillies

480 ml cider vinegar

675 g light soft brown sugar

100 g raisins

1 tsp ground allspice

Makes about 1.8 kg

1 ◀ Cut each mango on both sides of the stone; slice the flesh away from the stone and discard the stone. Peel the mango slices and cut the flesh into chunks. Put the mango flesh into a non-metallic bowl, stir in the salt, cover, and leave for 2 hours.

2 Put the tamarind pulp into a bowl and pour over boiling water to cover the pulp. Cover and leave for 1 hour, then drain off the excess water.

3 Put the tamarind pulp into a plastic sieve set over a bowl. Press the pulp through the sieve; discard the seeds. Drain the mangoes. Rinse the mango chunks in a plastic sieve under cold running water, until all the salt is washed off. Drain well. Put the sieved tamarind pulp and the mango flesh into a preserving pan.

4 Peel and finely chop the ginger. Halve the chillies lengthwise; remove and discard all the seeds. Either tear the chillies into small pieces with your fingers, or chop them with a small knife.

MYSTERY MANGO
A delicious complement to curries, simple cheese platters, and hot or cold meats.

5 ◀ Stir the ginger, chillies, vinegar, sugar, raisins, and allspice into the mango and tamarind mixture. Bring to a boil, stirring. Simmer over a medium heat, stirring, for 30 minutes, or until the mangoes are tender and the chutney has reduced and thickened.

6 ▶ Test by pulling the back of the spoon across the bottom of the pan. There should be no runny vinegar visible.

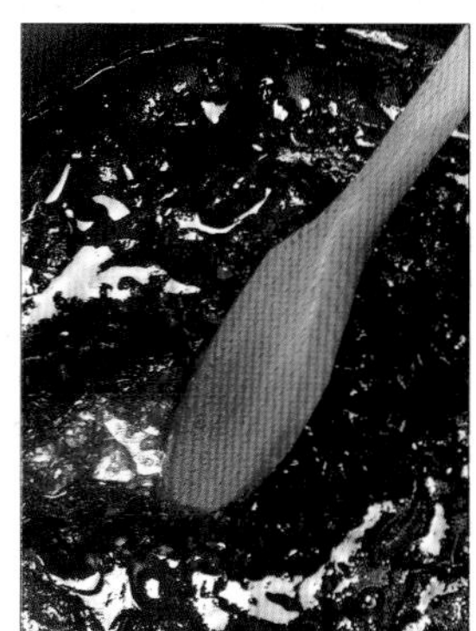

7 Spoon the chutney into warmed sterilized jars, to within 3 mm of the tops. Stir, if necessary, to remove any air pockets. Seal the jars and label. The chutney is now ready to use.

Autumn Fruit Chutney

INGREDIENTS

450 g pears

450 g cooking apples

450 g red plums

225 g onions

225 g raisins

600 ml cider vinegar

finely grated zest and juice of 1 orange

350 g brown sugar

1/2 tsp ground allspice

Making Autumn Fruit Chutney is the ideal way to use up end-of-season fruits. Any variety of pears and plums can be used, or they can be substituted by quinces and apricots.

1 Core the pears and apples, without peeling them. Cut them into small chunks and put them into a preserving pan.

2 Stone the plums, and peel and chop the onions. Add them to the pan with the raisins, vinegar, and orange zest and juice.

3 Bring the mixture to a boil, stirring. Simmer over a low heat, stirring occasionally, for about 45 minutes.

4 Add the sugar and allspice and stir over a low heat until the sugar has completely dissolved. Simmer, stirring occasionally, for 1 hour, or until the chutney has reduced and thickened.

5 Spoon the chutney into warmed sterilized jars, to within 3 mm of the tops. Seal the jars and label. Keep in a cool dark place for 2 months before using, to allow the flavours to develop.

Makes about 2 kg

Nectarine Chutney

INGREDIENTS

675 g firm nectarines

225 g onions

2 garlic cloves

50 g fresh root ginger

225 g sultanas

1 cinnamon stick

1 dried red chilli, crushed

2 tsp salt

1/2 tsp freshly grated nutmeg

240 ml orange juice

350 ml cider or white wine vinegar

1 Halve the unpeeled nectarines and remove the stones. Chop the flesh very coarsely. Peel and chop the onions and garlic. Peel and finely chop the root ginger.

2 Put the prepared ingredients into a large saucepan with all the remaining ingredients. Bring to a boil, stirring. Simmer over a low heat, stirring occasionally, for 40 minutes, or until the chutney has reduced and thickened. Discard the cinnamon stick.

3 Spoon the chutney into warmed sterilized jars, to within 3 mm of the tops. Seal the jars and label. The chutney is now ready to use.

Makes about 1.1 kg

DID YOU KNOW? There are more than 100 varieties of nectarines, with new varieties being developed every year. The flesh is very juicy with a good balance of sweetness of acidity, making it ideal for use in chutneys.

DATE AND ORANGE CHUTNEY

INGREDIENTS

450 g dried dates

450 g oranges

450 g onions

675 g sugar

75 g golden syrup

1 1/2 tbsp sea salt

1/4 tsp crushed dried red chillies

1.4 litres malt vinegar

450 g raisins

1 Stone and chop the dates. Finely grate the zest of 2 oranges, and set aside. Remove and discard the peel, pith, and pips from all the oranges. Chop the orange flesh. Peel and chop the onions.

2 Combine the sugar, golden syrup, salt, chillies, and vinegar in a preserving pan. Simmer over a low heat, stirring frequently, until the sugar has completely dissolved. Bring to a boil.

3 Add the dates, oranges, onions, raisins, and the grated orange zest. Lower the heat and simmer, stirring occasionally, for 1 hour, or until the chutney has reduced and thickened.

4 Spoon the chutney into warmed sterilized jars, to within 3 mm of the tops. Seal the jars and label. Keep in a cool dark place for 2 months before using, to allow the flavours to develop.

Makes about 2.5 kg

COOK'S TIP For a stronger orange flavour, grate the zest from all the oranges, add half during the cooking, and the other half just before spooning the chutney into the jars.

PINEAPPLE CHUTNEY

INGREDIENTS

2 pineapples with leaves removed, total weight about 1 kg

175 g sugar

100 ml white wine vinegar

2 tsp curry powder

1 tsp ground cinnamon

1/2 tsp ground cloves

1/2 tsp ground ginger

Lightly spiced and fruity, this chutney is ideal for serving with grilled ham, deep-fried Camembert, or smoked fish.

1 With a sharp knife, cut off the peel from the pineapple in strips, cutting deep enough to remove the "eyes" with the peel. Cut the pineapples across into thick slices, and cut out and discard the hard central core from each slice; dice the flesh.

2 Put the sugar, vinegar, spices, and any juice from the pineapple into a preserving pan.

3 Simmer over a low heat until the sugar has completely dissolved. Bring to a boil. Simmer for 5 minutes, or until the syrup reduces in volume by two-thirds.

4 Add the pineapple pieces and simmer over a low heat, stirring occasionally, for 20 minutes, or until the pineapple is soft.

5 Spoon the chutney into warmed sterilized jars, to within 3 mm of the tops. Seal the jars and label. The chutney is now ready to use.

Makes about 675 g

RED TOMATO CHUTNEY

A mild and aromatic fruit and vegetable chutney that is made with a careful balance of spices. Let the flavours blend and mature by storing the chutney for a couple of months before eating. It is the perfect accompaniment to thick wedges of mature cheese and crusty fresh bread.

INGREDIENTS

1.4 kg ripe tomatoes

675 g small onions

1 kg cooking apples

480 ml white wine vinegar

350 g sugar

175 g sultanas

2 tsp salt

1 tsp ground cloves

1 tsp ground ginger

1/2 tsp cayenne pepper

Makes about 2 kg

1 ◄ Cut the cores out of the tomatoes. Put the tomatoes into a bowl and cover with boiling water. Leave for 15–20 seconds, or until the skins start to split. Transfer the tomatoes to a bowl of cold water. Remove from the water one at a time and peel away the skins, using a sharp knife. Roughly chop the tomatoes.

2 ► Peel and thinly slice the onions. Peel, core, and chop the apples. Put the tomatoes, onions, and apples into a preserving pan. Add all the remaining ingredients and stir to combine.

3 ◄ Bring to a boil, stirring. Lower the heat and simmer, stirring often, for 40–45 minutes, or until the fruit and vegetables are soft and the chutney has reduced and thickened. Test by drawing the back of the spoon across the bottom of the pan. There should be no runny liquid visible.

4 Spoon the chutney into warmed sterilized jars, to within 3 mm of the tops. Stir to remove any air pockets. Seal the jars and label. Keep in a cool dark place for 2 months before using, to allow the flavours to develop.

Lemon and Mustard Seed Chutney

An excellent accompaniment for a Cheddar cheese ploughman's platter. If the mustard seeds are omitted, the plain lemon chutney tastes good with Thai or Indian food.

INGREDIENTS

8 large lemons, total weight about 1.8 kg (4 lb)

450 g onions

50 g salt

450 ml water

1 kg sugar

225 g sultanas

25 g mustard seeds

2 tsp ground ginger

1 tsp cayenne pepper

900 ml cider vinegar

1 Cut the lemons into 3 mm slices. Cut the slices into quarters; remove and discard any pips. Peel and slice the onions.

2 Layer the lemon and onion slices with the salt in a large non-metallic bowl. Put a plate on top to keep the mixture lightly pressed down. Leave, covered, for 24 hours, to draw out the excess moisture.

3 Rinse the lemon and onion slices in a plastic sieve or colander under cold running water until all the salt is washed off. Put them into a preserving pan and pour in water to cover. Bring to a boil. Simmer over a low heat, stirring occasionally, for 35 minutes, or until the lemon peel is very tender.

4 Add the remaining ingredients and return to a boil, stirring. Simmer over a low heat, stirring occasionally, for 45–50 minutes, or until the chutney has reduced and thickened.

5 Spoon the chutney into warmed sterilized jars, to within 3 mm of the tops. Seal the jars and label. Keep in a cool dark place for 2 months before using, to allow the flavours to develop.

Makes about 2.25 kg

Prune and Hazelnut Chutney

This fruity chutney does not need a long time to mature, in fact, it seems to have a better flavour when it is young. Serve with chicken or ham.

INGREDIENTS

1 kg prunes

675 g cooking apples

3 tbsp water

575 g brown sugar

225 g hazelnuts

600 ml red wine vinegar

1 tsp curry powder

1 tsp ground cinnamon

1/2 tsp ground allspice

1/4 tsp cayenne pepper

1 Put the prunes into a bowl and cover with boiling water. Leave to stand for 24 hours.

2 Peel, core, and finely chop the apples. Put them into a preserving pan with the water and 50 g of the sugar. Stir over a low heat until the sugar has completely dissolved. Simmer for 10 minutes, or until the apples are tender.

3 Chop the nuts finely. Drain the prunes and remove their stones. Chop the prunes and add to the pan with the nuts and all the remaining ingredients. Stir to combine.

4 Bring to a boil, stirring. Simmer over a low heat, stirring frequently, for 30 minutes, or until the chutney has reduced and thickened.

5 Spoon the chutney into warmed sterilized jars, to within 3 mm of the tops. Seal the jars and label. Keep in a cool dark place for 6 weeks before using, to allow the flavours to develop.

Makes about 2.25 kg

BANANA CHUTNEY

INGREDIENTS

450 g bananas

225 g dried dates

50 g crystallized ginger

450 ml white wine vinegar

finely grated zest and juice of 1 lemon and 1 orange

225 g raisins

275 g brown sugar

2 tsp each of salt and curry powder

1 Peel the bananas and cut the flesh into small pieces. Stone and chop the dates, and chop the crystallized ginger.

2 Put all the prepared ingredients into a preserving pan with the vinegar, lemon and orange zest and juice, and bring to a boil. Lower the heat and simmer, stirring occasionally with a wooden spoon, for 30 minutes.

3 Add all the remaining ingredients and stir over a low heat until the sugar has completely dissolved. Return to a boil and simmer, stirring frequently, for a further 10–15 minutes, or until the chutney has reduced and thickened.

4 Spoon the chutney into warmed sterilized jars, to within 3 mm of the tops. Seal the jars and label. Keep in a cool dark place for 1–2 months before using, to allow the flavours to develop.

Makes about 1.15 kg

PAWPAW CHUTNEY

INGREDIENTS

1 kg unripe pawpaw

100 g cashew nuts

50 g fresh root ginger

3 large garlic cloves

2 fresh hot red chillies

225 g sultanas

450 g brown sugar

480 ml cider vinegar

2 tsp salt

1 Peel the pawpaw, scrape out and discard the black seeds, and cut the flesh into 2.5 cm cubes.

2 Toast the cashew nuts on a baking tray in a 180°C (350°F) oven until lightly browned. Let them cool slightly, and coarsely chop. Peel and finely chop the root ginger and garlic cloves. Halve, seed, and chop the chillies.

3 Put all the ingredients into a preserving pan. Simmer, stirring occasionally with a wooden spoon, for 30 minutes, or until the chutney has reduced and thickened.

4 Spoon the chutney into warmed sterilized jars, to within 3 mm of the tops. Seal the jars and label. Keep in a cool dark place. The chutney can be used immediately, but it improves if stored for 2 weeks before using, to allow the flavours to develop.

Makes about 1.25 kg

ORANGE CHUTNEY

INGREDIENTS

6 oranges, total weight about 1 kg

225 g onions

450 g fresh dates

600 ml white wine vinegar

2 tsp each of ground ginger and coriander

1 Peel the oranges, discarding the peel and pips. Chop the flesh, and put it, with any juice, into a preserving pan.

2 Peel and chop the onions, and stone and chop the dates. Add to the pan with the oranges. Add all the remaining ingredients and bring to a boil. Simmer, stirring occasionally, for 1 hour, or until the chutney has reduced and thickened.

3 Spoon into warmed sterilized jars, to within 3 mm of the tops. Seal the jars and label. Keep in a cool dark place for $1\frac{1}{2}$–2 months before using, to allow the flavours to develop.

Makes about 1.1 kg

Beetroot Chutney

INGREDIENTS

2 cooking apples

1 onion

240 ml malt vinegar

2 tsp freshly grated root ginger

1 tsp ground allspice

2 whole cloves

450 g cooked beetroots (see box, right)

50 g brown sugar

50 g raisins

1 Core and slice the apples. Peel and slice the onion. Put the apples and onion into a preserving pan. Add the vinegar with the root ginger, allspice, and cloves; bring to a boil. Simmer, stirring occasionally, for 20 minutes.

2 Peel the cooked beetroots, finely chop them, and add to the pan with the sugar and raisins. Bring to a boil and simmer for 15 minutes.

3 Spoon the chutney into warmed sterilized jars, to within 3 mm of the tops. Seal the jars and label. The chutney is now ready to use.

Makes about 1 kg (2 lb)

TO COOK BEETROOTS Do not peel them before cooking because they will "bleed". Bring a pan of salted water to a boil, add beetroots, and simmer for 20–30 minutes. Drain and cool.

Bengal Chutney

INGREDIENTS

225 g carrots

1 onion

450 g cooking apples

100 g raisins

100 g freshly grated horseradish

225 g demerara sugar

2 tsp salt

3 tsp ground ginger

3 tsp curry powder

1 tsp mustard seeds

1½ tbsp golden syrup

300 ml vinegar

Chutneys are served as condiments in an Indian meal in the same way as salt and pepper are in the West. This chutney comes from Bengal, in northern India.

1 Peel and slice the carrots and onion. Peel, core, and slice the apples. Put all the sliced ingredients into a saucepan.

2 Add all the remaining ingredients to the pan. Simmer, stirring occasionally, for 50 minutes, or until the chutney has reduced and thickened.

3 Spoon the chutney into warmed sterilized jars, to within 3 mm of the tops. Seal the jars and label. Keep in a cool dark place for 6 weeks before using, to allow the flavours to develop.

Makes about 1.15 kg

Pear and Onion Chutney

INGREDIENTS

1.4 kg pears

675 g onions

450 g ripe tomatoes

2 small green capsicums

4 tbsp raisins

450 g brown sugar

3 tsp salt

1/2 tsp ground cinnamon

1/4 tsp ground cloves

pinch of cayenne pepper

900 ml white wine vinegar

It may seem unusual to mix pears, onions, tomatoes, and capsicums but this unlikely combination works surprisingly well. As with most chutneys, it is a delicious accompaniment to curries, Thai food, and a traditional ploughman's lunch.

1 Peel, core, and dice the pears. Peel and slice the onions, and peel and chop the tomatoes. Core, seed, and chop the capsicums. Discard all the peel, cores, and seeds.

2 Put all the prepared vegetables into a preserving pan, and simmer over a low heat for 20–30 minutes, or until the vegetables are soft.

3 Add all the remaining ingredients and slowly bring to a boil, stirring with a wooden spoon, until the sugar has completely dissolved. Simmer, stirring occasionally, for 1–1 1/2 hours, or until the chutney has reduced and thickened.

4 Spoon the chutney into warmed sterilized jars, to within 3 mm of the tops. Seal the jars and label. Keep in a cool dark place for 2 months before using, to allow the flavours to develop.

Makes about 2.5 kg

Rhubarb Chutney

INGREDIENTS

450 g rhubarb

350 g onions

175 g raisins

775 g brown sugar

480 ml cider vinegar

3 tsp salt

1 tsp ground cinnamon

1 tsp ground ginger

1/2 tsp ground cloves

pinch of cayenne pepper

1 Trim and slice the rhubarb. Peel and slice the onions.

2 Put the rhubarb and onions into a stainless steel preserving pan with all the remaining ingredients.

3 Stir over a low heat, with a wooden spoon, until the sugar has completely dissolved. Bring to a boil. Simmer, stirring frequently, for 2 hours, or until the chutney has reduced and thickened.

4 Spoon the chutney into warmed sterilized jars, to within 3 mm of the tops. Seal the jars and label. Keep in a cool dark place for 1 month before using, to allow the flavours to develop.

Makes about 2.25 kg

COOK'S TIP Always prepare rhubarb properly before cooking – top and tail the rhubarb, and peel off any tough strings if necessary. Be sure to remove all the leaves, which are poisonous and should never be eaten. Always use a stainless steel pan when cooking rhubarb, because the acidity of the rhubarb may react with other metals.

SPICED CAPSICUM CHUTNEY

INGREDIENTS

6 red capsicums

6 green capsicums

1 kg green or unripe tomatoes

675 g onions

675 g cooking apples

1 tsp ground allspice

900 ml cider vinegar

450 g sugar

25 g salt

10 whole dried chillies

1 tsp whole cloves

7 g finely grated fresh root ginger

15 g mustard seeds

25 g peppercorns

This chutney is both colourful and flavourful when served as an accompaniment to a cheese platter.

1 Halve the capsicums and the tomatoes, and remove and discard the cores and seeds. Peel the onions and peel and core the apples. Coarsely chop all the vegetables and the apples in a food processor, or chop them roughly with a large knife.

2 Put the vegetables and fruit into a preserving pan with the allspice, vinegar, sugar, and salt. Add all the remaining ingredients, tied together tightly in a muslin bag. Stir over a low heat, with a wooden spoon, until the sugar has completely dissolved. Simmer, stirring occasionally, for 1¼ hours, or until the chutney has reduced and thickened.

3 Remove the muslin bag, and spoon the chutney into warmed sterilized jars, to within 3 mm of the tops. Seal the jars and label. Keep in a cool dark place for 6 weeks before using, to allow the flavours to develop.

Makes about 2.75 kg

DID YOU KNOW? Sweet capsicums are mild in flavour and can be eaten cooked or raw. Chillies or hot capsicums are used as a seasoning, but don't be fooled by the small green ones, which are often more fiery than red chillies.

APRICOT AND ALMOND CHUTNEY

INGREDIENTS

240 ml cider vinegar

175 g sugar

12 apricots

2 red capsicums

2 onions

1 garlic clove

1 orange

1 lemon

50 g crystallized ginger

1 tsp salt

75 g raisins

50 g whole blanched almonds

1 tsp ground ginger

1 Pour 175 ml of the vinegar into a preserving pan. Add the sugar. Stir over a low heat until the sugar has completely dissolved. Increase the heat and bring the mixture to a boil. Simmer for 5 minutes.

2 Halve, stone, and chop the apricots. Core, seed, and chop the capsicums. Peel and chop the onions and garlic. Finely chop the whole orange and lemon, including the peel and pith. Finely chop the crystallized ginger.

3 Add the prepared fruits and vegetables to the vinegar mixture together with the crystallized ginger, salt, and raisins. Simmer over a medium heat, stirring frequently, for 30 minutes.

4 Add the almonds, ground ginger, and remaining vinegar. Simmer for a further 30 minutes, stirring frequently, or until the chutney has reduced and thickened.

5 Spoon the chutney into warmed sterilized jars, to within 3 mm of the tops. Seal the jars and label. Keep in a cool dark place for 2 months before using, to allow the flavours to develop.

Makes about 1.4 kg

RELISHES

WITH THEIR SPICY TANG and crisp texture, relishes provide a challenge to the cook who can choose either to follow the excellent example of tradition, or to abandon the past and make creative forays into new culinary territory. Either way, the results are bound to embellish the dishes they supplement. Fine relishes are as important as a fine sauce, and deserve as much attention. They are usually combinations of fruits, sweet or tart, and crisp vegetables, such as celery, radishes, carrots, and sweet capsicums, and they complement both hot and cold meats, fish, and rich poultry, especially duck and goose. Relishes are delicious with cottage cheese or ricotta as an appetizer or a light luncheon dish. They can be stirred into mayonnaise for a salad dressing, they are good in sandwiches, and give meat loaves an extra sweet-sour or hot chilli flavour.

Most ingredients for relishes are available all year round, with the exception of that great favourite, Sweetcorn Relish, which has to be made with kernels from fresh corn cobs during the brief sweetcorn season. There is a temptation to sample relishes to oblivion: just a taste here, a taste there, and the jar is empty. Always make enough for the storecupboard or refrigerator, and plenty for gift-giving.

MAKING RELISHES

- Choose young fresh fruits or vegetables. If using vegetables with a high water content, such as cabbages, cucumbers, and sweet capsicums, layer them with sea salt and leave overnight to draw out the moisture.
- Drain the vegetables and rinse under cold running water to wash off the extra salt, drain again, and pat dry with kitchen paper.
- Cook the relish in a preserving pan with your choice of vinegar, sugar, and spices. The fruits and vegetables are cooked until they are just tender. Relishes are distinct from chutneys in that they are more coarsely textured with a sharper flavour, while chutneys are characteristically smooth and mellow. For this reason, relishes require less cooking than chutneys.
- Spoon the relish into warmed sterilized jars (see page 11), to within 3 mm of the tops,and stir, if necessary, to remove any air pockets.

How to seal and store

Seal the jars very tightly with non-corrosive, screw-top lids and label. Store in a cool dark place for 2–4 weeks before using, to allow the flavours to develop. Label the jars. The relishes should then keep for 3 months, depending on the ingredients used. It is a good idea to date the relishes so that it is clear when they should be used up. Once the jars of relish have been opened, keep them in the refrigerator.

What can go wrong and why

If the relish shrinks in the jar, the seal is not airtight. Relishes will lack flavour and a sharp taste if they are not left to mature for long enough before using, or if a poor-quality vinegar has been used. If the colour of the relish fades, it has been exposed to light, which hastens oxidation and destroys certain vitamins.

SWEETCORN RELISH

In the height of summer, when gardens and market stalls are overflowing with fresh ears of corn, it is time to make this traditional tangy relish. It can be eaten straight away, or tucked into a cupboard to enjoy in the winter months.

INGREDIENTS

8 fresh ears of corn, total weight about 2.5 kg

2 each of red capsicums, green capsicums, and onions

8 celery sticks

1.15 litres cider vinegar

100 g sugar

3 tsp mustard seeds

3 tsp salt

4 allspice berries

Makes about 2.5 kg

1 ▶ Remove and discard the husks and silks from the fresh ears of corn.

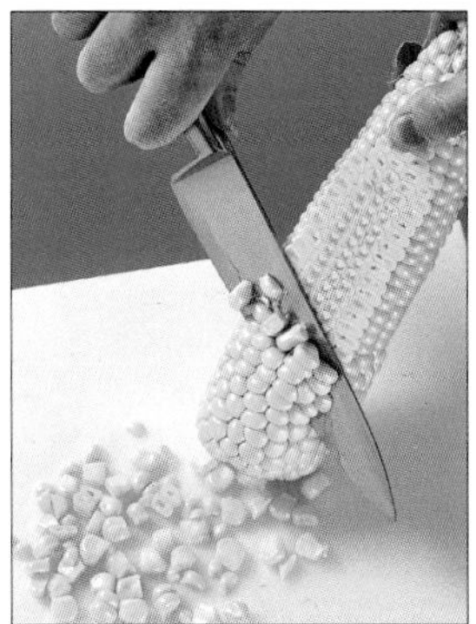

2 ◀ Cut the corn kernels from each cob. Cut around the core of each capsicum, and pull it out. Halve each capsicum lengthwise and scrape out the seeds. Cut away the white ribs on the inside. Cut each capsicum half lengthwise into strips and then crosswise into dice.

3 ▶ Peel and finely chop the onions. Finely chop the celery. Put all the vegetables into a preserving pan. Add all the remaining ingredients to the pan, and stir over a low heat until the sugar has completely dissolved.

4 ◀ Bring the mixture to a boil, stirring. Simmer stirring occasionally, for 15–20 minutes, or until the vegetables are tender.

5 ▲ Spoon the relish into warmed sterilized jars, to within 3 mm of the tops. Seal the jars and label.

Horseradish Relish

INGREDIENTS

1.8 kg tomatoes

175 g celery

4 onions

225 g grated horseradish

350 g brown sugar

600 ml cider vinegar

1½ tbsp salt

2 tsp each of ground allspice, cinnamon, and cloves

4 tbsp mustard seeds

1½ tbsp dill seeds

1 Peel and chop the tomatoes. Put into a plastic sieve and leave to drain for 2 hours. Chop the celery, and peel and chop the onions.

2 Put the prepared vegetables into a heavy-based saucepan. Add all the remaining ingredients. Stir over a low heat until the sugar has completely dissolved. Bring to a boil and simmer, stirring occasionally, for 50 minutes, or until the vegetables are tender. Spoon the relish into warmed sterilized jars, seal, and label. Keep in a cool dark place for 2 months before using, to allow the flavours to develop.

Makes about 2.25 kg

COOK'S TIP To prepare freshly grated horseradish, peel the skin down to the flesh. Grate in a food processor (using a hand grater can produce very strong fumes, causing burning and watering eyes). Dried horseradish flakes can be reconstituted in water and used as fresh.

Jerusalem Artichoke Relish

INGREDIENTS

2 kg Jerusalem artichokes

4 onions

2 green or red sweet capsicums

3 tsp sea salt

1 tsp each of dill and mustard seeds, and turmeric

1.4 litres cider vinegar

575 g sugar

1 Peel the artichokes and onions. Core and seed the capsicums. Coarsely mince the vegetables in a food processor, then put into a preserving pan. Add all the remaining ingredients and stir over a low heat until the sugar has completely dissolved. Bring to a boil, and simmer, stirring occasionally, for 30 minutes or until the vegetables are tender.

2 Spoon the relish into warmed sterilized jars. Seal the jars and label. Keep in a cool dark place for 1 week before using, to allow the flavours to develop.

Makes about 3 kg

DID YOU KNOW? The Jerusalem artichoke bears no relation to the globe artichoke, but is in fact a member of the sunflower family. It has a rather odd shape and texture, but its flavour is sweet and nutty.

Cranberry-Orange Relish

INGREDIENTS

450 g frozen cranberries

225 g sugar

2 oranges

Unlike other relishes, this relish does not need cooking, so it is quick and easy to prepare.

1 Chop the cranberries and put them into a non-metallic bowl with the sugar. Stir to mix well together.

2 Slice the unpeeled oranges and remove the pips. Finely chop the orange slices in a food processor, and add them to the bowl of cranberries and sugar.

3 Spoon the relish into sterilized jars. Seal, label, and refrigerate immediately. The relish should be chilled for several hours before serving. It will keep for about 1 week in the refrigerator.

Makes about 1 kg

CARROT AND CUCUMBER RELISH

INGREDIENTS

225 g carrots

225 g onions

675 g cucumber

1 green capsicum

1 red capsicum

2 small fresh green chillies

175 g sea salt

200 ml cider vinegar

150 g sugar

1 tsp mustard seeds

1 tsp fennel seeds

This appetizing relish is a great favourite. It is good with any cold meats or chicken, hamburgers, barbecued meats, or quiches. Once the jars have been opened, keep them in the refrigerator.

1 Peel and coarsely grate the carrots. Peel and finely chop the onions. Finely dice the cucumbers. Core, seed, and chop the capsicums and chillies.

2 Layer the vegetables with the salt in a large non-metallic bowl. Leave to stand overnight, to draw out excess moisture.

3 Drain the vegetables and rinse in a plastic sieve or colander under cold running water, until all the salt is washed off. Drain again and dry on kitchen paper.

4 Combine all the remaining ingredients in a saucepan and bring to a boil. Add the drained vegetables and simmer for 10 minutes, or until the vegetables are tender.

5 Spoon the relish into warmed sterilized jars. Seal the jars and label. Keep in a cool dark place for 2 weeks before using, to allow the flavours to develop.

Makes about 1.15 kg

END-OF-SEASON RELISH

INGREDIENTS

450 g green or unripe tomatoes

225 g ripe tomatoes

1/2 small green cabbage

2 red capsicums

2 green capsicums

2 sticks celery

2 onions

1/2 cucumber

50 g sea salt

225 g brown sugar

900 ml cider or white wine vinegar

This is an excellent relish for using up excess garden produce left at the end of the summer. It is good with meat and poultry, and with a sharply flavoured cheese, such as a mature Cheddar.

1 Chop the tomatoes and cabbage coarsely. Cut around the core of each capsicum and pull it out. Halve each capsicum lengthways and scrape out the seeds. Cut away the white ribs on the inside, and dice the capsicums coarsely. Chop the celery, and peel and chop onions. Peel and chop the cucumber.

2 Layer the vegetables with the salt in a large non-metallic bowl. Leave to stand overnight, to draw out excess moisture.

3 Drain the vegetables, pressing them to remove as much liquid as possible. Rinse in a plastic sieve under cold running water, drain again, and pat dry on kitchen paper. Put the vegetables into a large saucepan. Add the sugar and vinegar; stir to mix. Simmer the mixture for 1 hour, or until the vegetables are tender.

4 Spoon the relish into warmed sterilized jars. Seal the jars and label. Keep in a cool dark place for 2 weeks before using, to allow the flavours to develop.

Makes about 2 kg

MUSTARDS

MUSTARD HAS BEEN cultivated since 4,000 BC; it is a versatile plant and is one of the most popular food flavourings in the world. There are huge numbers of mustard recipes, all based on three types of seed, the *nigra* (black), *juncea* (brown), and *alba* (white or yellow). Mustard seasoned the food of the ancient Egyptians, Greeks, and Romans, and is as prolific as it is ancient: a seed of black mustard can produce 3,400 young plants, making it a Hindu symbol of fertility. The name comes from the Latin *mustum ardens* or "burning must", which refers to the pungent taste that develops when the seeds are ground with must, or unfermented grape juice. Back in AD 800, the Emperor Charlemagne noted that the leaves of the mustard plant could be eaten either raw or cooked; today it is a popular vegetable in India, China, Japan, Africa, and parts of the USA. Mustard turns up in every type of cuisine from appetizers, soups, sauces, and pickles, through fish, poultry, game and meat, to vegetables and salads, and even desserts and puddings.

Mustard is a member of the cabbage family, and its seeds are classified as mild to strong. The simplest form of mustard is dry mustard powder, which is a mix of ground seeds, very hot, and a favourite in England, China, and Japan. There are also whole-grain mustards, and speciality mustards like Dijon, German, American, Beaune, Bordeaux, and Beaujolais, that vary in taste because of the proportions of seeds ground with vinegar, sugar, and spices. Mustard comes with many different flavourings, such as herbs, lemon, chilli, and honey, to name just a few. Each flavouring gives a delectably different taste to the mustard.

MAKING MUSTARDS

❦ Select the type of seeds and flavourings. Basic mustards can be mild or strong, smooth or coarse, and the variety of seeds you use will determine the taste. White or yellow seeds are mild, black seeds are strong and pungent, and brown seeds are hot and aromatic. If you want to make a flavoured mustard, use commercially ground dry mustard powder as a base for a really smooth texture. Alternatively, for a rougher texture, grind your own seeds in an electric grinder or in a mortar and pestle.

❦ Add the flavourings, either aromatic fresh herbs, such as tarragon and basil, hot spices like green peppercorns and crushed dried chillies, or fruits, from sharp citrus to sweet berries. Even freshly grated root ginger or horseradish can be used. A sweetener, such as caster sugar or honey, can also be added. Vinegar is usually used to blend the ingredients together. Whisky or beer can also be stirred in.

❦ Spoon the mustard into sterilized jars (see page 11).

Traditional mustard pots have small necks; these keep mustard fresh for a long time, because they prevent a large surface area from coming into contact with the air.

How to seal and store

Seal the jars tightly with non-corrosive, screw-top lids, and keep in a cool dark place for 1–2 weeks before using, to allow the flavours to develop. Use the mustard within 3 months. Once opened, the flavour will start to deteriorate, so refrigerate and use it up fairly quickly.

What can go wrong and why

If the mustard dries out on the surface, it has not been sealed correctly. If the mustard is left too long, it will lose most of its strength.

Whole-Grain Mustard

INGREDIENTS

100 g whole black or brown mustard seeds

175 ml–200 ml white wine vinegar

50 g whole yellow mustard seeds

3 tsp salt

1 Put the black mustard seeds into a non-metallic bowl and pour over 150 ml of the vinegar. Cover and leave overnight.

2 The next day, use a mortar and pestle to pound the mixture until the seeds are coarsely broken. Grind the yellow mustard seeds in an electric blender to a very fine powder. Combine the two mixtures and stir in the rest of the vinegar and the salt.

3 Spoon into sterilized jars. Seal, label, and keep in a cool dark place for 2 weeks before using, to allow the flavours to develop.

Fills about three 150 ml jars

Tarragon Mustard

INGREDIENTS

15 g fresh tarragon

50 g plain flour

75 g mustard powder

25 g caster sugar

1 tbsp salt

100 ml white wine vinegar

1 Strip the tarragon leaves from their stalks. Discard the stalks and finely chop the leaves.

2 Sift the flour into a bowl, and add the mustard powder, caster sugar, and salt. Mix the ingredients together well. Add the tarragon and vinegar to the bowl, and stir the mixture until a smooth paste forms.

3 Spoon the mustard into sterilized jars. Seal the jars and label. Keep in a cool dark place for 1 week before using, to allow the flavours to develop.

Fills about three 75 ml jars

VARIATIONS

Other flavourings can enhance the taste of mustards. Omit the fresh tarragon in this recipe and add different herbs, such as basil, mint, parsley, sage, thyme, or rosemary.

English Mustard

INGREDIENTS

100 g whole yellow mustard seeds

15 g plain flour

3 tsp salt

175 ml light beer

1 Put the mustard seeds in an electric grinder and grind to a fine powder. Transfer to a bowl, and sift in the flour and salt. Mix together well. Gradually beat in the beer to make a smooth paste.

2 Spoon the mustard into sterilized jars. Seal the jars and label. Keep in a cool dark place for 2 weeks before using, to allow the flavours to develop.

Fills about two 150 ml jars

Horseradish Mustard

INGREDIENTS

50 g plain flour

75 g mustard powder

25 g caster sugar

1 tbsp salt

75 g fresh horseradish

175 ml cider vinegar

1 Sift the flour into a bowl, add the remaining dry ingredients, and mix together well.

2 Peel and finely grate the horseradish. Add to the dry ingredients with the vinegar and mix again to make a smooth paste.

3 Spoon the mustard into sterilized jars. Seal the jars and label. Keep in a cool dark place for 1 week before using, to allow the flavours to develop.

Fills about four 75 ml jars

Luxurious Oils & Vinegars

Flavoured Oils

Infusing oils with a single flavour or a compatible combination of herbs and spices is a lovely way to make personalized gifts for friends. Almost any herb or spice can be used, along with aromatic fruits, such as oranges, lemons, or limes. Flowering herbs, such as rosemary and thyme, added to the finished bottles create a pretty effect.

Today's cooks have a wide choice of oils to work from, each with a special flavour of its own. Everyday oils – corn, rapeseed (canola), safflower, and sunflower – are guaranteed not to dominate the taste of the ingredients with which they are combined; these oils are generally suitable for infusing with the more pungent herbs and spices. Luxury oils, including walnut and hazelnut, and the exotics, such as mustard oil used in Indian cooking, and the bright orange, nutty palm oil used in Brazilian cooking – already have their own distinctive tastes and are not usually chosen for making flavoured oils. The greatest of all the oils for flavouring is olive, and the best quality is extra-virgin olive oil, made from the first cold pressing of the finest olives.

Flavoured oils can be used in marinades, salad dressings, or for the initial sauté in a stir-fry; in fact, for any dish in which you would use an ordinary oil. Simply ensure that the taste of the oil complements the ingredients you are using.

Making Flavoured Oils

- Start by sterilizing glass bottles (see page 11). It is best to choose a small size, because oils can turn rancid after opening and small amounts will be used more quickly.
- Select your flavourings. If you are using herbs, they should be freshly picked and lightly dried on kitchen paper to remove excess moisture, then bruised lightly to release their aroma immediately. Dried spices, such as cinnamon sticks, peppercorns, and chillies, should be as fresh as possible so that their maximum flavour is extracted. Fresh garlic cloves lend pungency, but should only be used in small amounts. If garlic is added to flavour the oil, it is important to keep the oil refrigerated and to use it within 2 weeks.
- Insert the prepared flavourings into the bottles, then pour in the oil to fill and cover all the ingredients (so mould cannot grow). Seal the bottles.
- Leave for about 2 weeks in a cool dark place. During this time the flavour will develop in the oil, so taste it occasionally to see when it is ready to use. Shake the bottle a few times, especially if it contains ingredients such as paprika that settle on the bottom. If you prefer the flavour not to get any stronger at this stage, or if you want the oil to last longer (particularly if it contains fresh ingredients), strain it through a double layer of muslin into freshly sterilized bottles. For decoration and easy identification, a fresh herb sprig can be added to herb oils, or a twist of orange or lemon zest to citrus oils.

How to seal and store

Seal tightly in bottles with non-corrosive, screw-top lids, label, then keep in the refrigerator for 3–6 months. The storage time will depend on the flavouring of the oil. Those that contain fresh ingredients, such as herbs or fruits, unless strained first, will only last about 3 months. Oils that have been infused with dried ingredients will last about 6 months.

What can go wrong and why

A flavoured oil will become cloudy if its flavouring, such as fresh onion, contains too much water. If this happens, use up the oil as soon as possible, because it can quickly become rancid. Other reasons for an oil turning rancid are incorrect storage, faulty sealing of bottles, and the oil coming into contact with direct sunlight and heat.

FIERY CHILLI OIL

Brilliant red chillies, infused in warm oil, produce a glowing colour and potent taste that can be diluted by using fewer chillies than suggested here, or by taking out the chilli seeds. Use chilli oil to sauté onions and garlic for stir-fries, sauces, casseroles, or soups, or mix it with vinegar to make a salad dressing. It will give every dish an instant burst of flavour.

INGREDIENTS

12 dried red chillies

480 ml rapeseed (canola) oil or corn oil

1 tbsp cayenne pepper

1½ tbsp roasted sesame oil

Makes about 480 ml

1 ◄ Finely chop the chillies. Put them into a medium saucepan and pour in the rapeseed or corn oil. Simmer over a very low heat for 10 minutes. (Don't let the oil get too hot; over-heating will spoil its keeping qualities.) Remove from the heat and leave to cool.

2 ◄ Stir in the cayenne pepper and the sesame oil. Cover the pan and leave for 12 hours, to allow the flavours to develop.

3 ► Line a funnel with a double layer of muslin and strain the flavoured oil through it into a sterilized bottle, to within 3 mm of the top.

4 If you like, add 2–3 whole dried red chillies to the bottle of oil. Seal the bottle and label. The oil is now ready to use.

IDENTIFICATION PARADE
Two or three dried chillies added to the bottle of oil will make it readily identifiable on the kitchen shelf.

COOK'S TIP When chillies are lightly cooked in oil, their flavour is released instantly, so the oil can be used right away rather than left for 1–2 weeks.

HUNGARIAN OIL

INGREDIENTS

2 tsp paprika

1 tsp turmeric

240 ml rapeseed (canola) oil or corn oil

This oil is perfect for sautéing the onions, garlic, and meat when making an authentic goulash, but it can also be used to sauté fish, shellfish, poultry, and vegetables.

1 Put the paprika, turmeric, and oil into a small heavy-based saucepan. Simmer over a low heat, stirring occasionally, for 1–5 minutes, or until the mixture turns a rich orange-red colour. As soon as the colour of the oil begins to develop, remove the pan from the heat and leave to cool.

2 Line a funnel with a double layer of muslin and strain the flavoured oil through it into a sterilized bottle, to within 3 mm of the top.

3 Seal the bottle and label. The oil is now ready to use.

Makes about 240 ml

DID YOU KNOW?
Paprika, which ranges from sweet to sharp in flavour, is the national spice of Hungary. It is used extensively in the cooking of savoury dishes, and it is the essential flavour in classic Hungarian goulash.

SWEET PAPRIKA OIL

INGREDIENTS

3 tbsp sweet paprika

1 litre extra-virgin olive oil

1 Divide the paprika equally between sterilized bottles: spoon it through a handmade paper funnel placed in the neck of each of the bottles.

2 Pour the extra-virgin olive oil into the bottles, to within 3 mm of the tops.

3 Seal the bottles and shake well. Keep in a cool dark place for 1 week before using, to allow the flavours to develop. Shake the bottles from time to time.

4 Line a funnel with a double layer of muslin and strain the flavoured oil through it into freshly sterilized bottles. Seal the bottles and label. The oil is now ready to use.

Makes about 1 litre

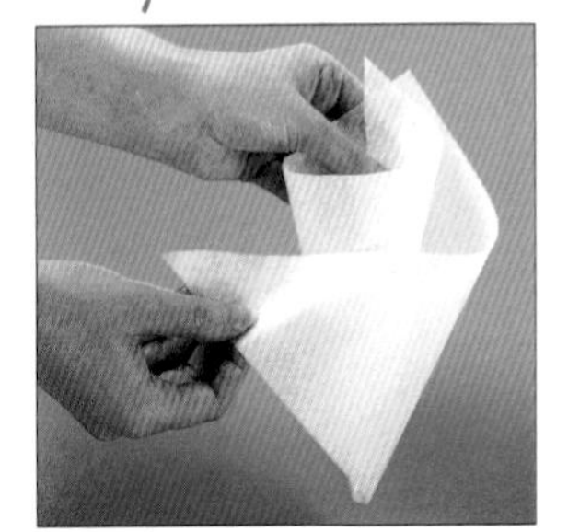

DO-IT-YOURSELF FUNNEL
A handmade funnel is useful for spooning ground spices into bottles. Make it with paper or foil.

ORANGE AND CORIANDER OIL

INGREDIENTS

1½ tbsp coriander seeds

4 strips of dried orange zest (see box, right)

1 litre extra-virgin olive oil

1 Lightly crush the coriander seeds in a mortar and pestle, being careful not to break them up completely. Alternatively, use the end of a rolling pin to crush them. Divide the coriander seeds equally between sterilized bottles.

2 If necessary, cut the dried orange zest to a size that will fit through the necks of the bottles. Add the zest to the coriander. Pour in the oil, to within 3 mm of the tops.

3 Seal the bottles and shake well. Label and keep in a cool dark place for 1 week before using, to allow the flavours to develop. Shake the bottles from time to time.

Makes about 1 litre

COOK'S TIP
To dry orange zest: preheat the oven on its lowest setting. Peel wide strips of zest from top to bottom of an orange. Put the strips on a baking sheet and leave in the oven for 1¼ hours, or until dry.

MIXED HERB OIL

INGREDIENTS

2 sprigs of fresh rosemary

2 sprigs of fresh thyme

2 pickling onions

2 bay leaves

12 black peppercorns

1 litre (1 3/4 pints) extra-virgin olive oil

1 Lightly bruise the rosemary and thyme to release their flavours. Peel and finely slice the pickling onions.

2 Divide the bruised herbs, onion slices, bay leaves, and peppercorns equally between 2 sterilized bottles.

3 Pour in the oil, to within 3 mm (1/8 inch) of the tops, making sure that the ingredients are completely covered.

4 Seal the bottles and shake well. Label and keep in a cool dark place for 2 weeks before using, to allow the flavours to develop. Shake the bottles from time to time.

Makes about 1 litre (1 3/4 pints)

PERFUMED THAI OIL

INGREDIENTS

4–6 sprigs of fresh coriander

6 pieces of fresh lemon grass

4 dried red chillies

1 litre (1 3/4 pints) rapeseed (canola) oil or corn oil

1 Lightly bruise the coriander and lemon grass with the flat side of a chef's knife, to release their flavours.

2 Divide the bruised coriander and lemon grass, and the chillies, equally between sterilized bottles. Pour in the oil, to within 3 mm (1/8 inch) of tops.

3 Seal the bottles and shake well. Label and keep in a cool dark place for 2 weeks before using, to allow the flavours to develop. Remove the coriander and lemon grass after 2 weeks.

Makes about 1 litre (1 3/4 pints)

INDIAN SPICE OIL

INGREDIENTS

5 ml (1 tsp) Aromatic Garam Masala (see recipe, page 125)

2.5 ml (1/2 tsp) garlic powder

2.5 ml (1/2 tsp) ground coriander

2.5 ml (1/2 tsp) ground cumin

2.5 ml (1/2 tsp) chilli powder

2.5 ml (1/2 tsp) ground turmeric

10 ml (2 tsp) dried fenugreek leaves

2 whole cloves

1 litre (1 3/4 pints) rapeseed (canola) oil or corn oil

This powerfully flavoured oil adds a tang to marinated or sautéed food, and is also ideal for barbecued steak and chicken.

1 Put all the powdered spices into a small bowl and mix well together. Divide the spice mixture equally between 2 sterilized bottles by spooning them through a funnel placed in each bottle neck. Divide the fenugreek leaves and cloves between the bottles.

2 Pour in the oil, to within 3 mm (1/8 inch) of the tops. Seal the bottles and shake well. Keep in a cool dark place for 1–2 weeks before using, to allow the flavours to develop. Shake the bottles from time to time.

3 Line a funnel with a double layer of muslin and strain the oil through it into 2 freshly sterilized bottles. Seal the bottles and label. The oil is now ready to use.

Makes about 1 litre (1 3/4 pints)

From left to right, *Perfumed Thai Oil, Sweet Paprika Oil, and Mixed Herb Oil*

OIL OF ITALY

INGREDIENTS

4 fresh sage leaves

4–6 sprigs of fresh basil

12 black peppercorns

1 litre extra-virgin olive oil

Italian cooks adore the earthy taste of sage and the semi-sweet flavour of basil, both of which are combined in this oil. Use it for tossing with pasta, or for drizzling over a mozzarella cheese and tomato salad.

1 Lightly bruise the sage and basil to release their flavours. Divide the bruised herbs and the peppercorns equally between sterilized bottles. Pour in the olive oil, to within 3 mm of the tops of the bottles.

2 Seal the bottles and shake well. Label and keep in a cool dark place for 2 weeks before using, to allow the flavours to develop. Shake the bottles from time to time.

Makes about 1 litre

PROVENÇAL OIL

INGREDIENTS

25 g mixed fresh herbs and spices, e.g. fennel stalks, oregano, tarragon, thyme, rosemary, coriander seeds, 1 bay leaf, and 1 dried red chilli

600 ml extra-virgin olive oil

An intriguing mix of herbs and spices goes into this oil, which takes on the scent of southern France as it soaks up their flavours. The herbs and spices listed here are just one suggestion, but you can experiment with other combinations.

1 Put the herbs and spices into a sterilized bottle. Pour in the olive oil, to within 3 mm of the top, making sure that the herbs are completely covered.

2 Seal the bottle and shake well. Label and keep in a cool dark place for 2 weeks before using, to allow the flavours to develop. Shake the bottle from time to time.

Makes about 600 ml

ROSEMARY OIL

INGREDIENTS

4–6 large sprigs of fresh rosemary

1 litre extra-virgin olive oil

This is the simplest of flavoured oils. Any other single herb can be used to flavour oil in the same way: lemon balm, tarragon, marjoram, oregano, mint, or thyme. For a stronger flavour, add extra herb sprigs and keep for longer before opening.

1 Lightly bruise the rosemary sprigs (see box, below), to release their flavour.

2 Divide the sprigs of rosemary equally between sterilized bottles. Pour in the olive oil, to within 3 mm of the tops, making sure the rosemary is completely covered.

3 Seal the bottles and shake well. Label and keep in a cool dark place for 2 weeks before using, to allow the flavours to develop. Shake the bottles from time to time.

Makes about 1 litre

DID YOU KNOW? To extract the full flavour from fresh herbs, it is best to bruise them just before using. Put the herbs on a chopping board and place a large chef's knife on top of them. Using the palm of your hand, press down firmly on the flat of the knife.

Preserves in Oil

Olive oil, richly aromatic and fruity, is wonderful as a preserver of foods. In the commercial world, artichoke hearts, anchovies, sardines, and tuna are all traditionally preserved in olive oil. We may take them for granted, yet those of us addicted to the sharp, salty flavour of anchovies in a Caesar salad, or chunks of tuna tossed in a *salade Niçoise*, would greatly miss these treats if there were no oil to preserve them for us.

There are many gifts to be made at home, using a variety of foods with oil as a preservative. Olives, mushrooms, feta and goat's cheeses, and lemons, will all be transformed once bathed and soaked in oil. Pack them in your favourite jars, jazz them up with fresh herbs and spices to bring out their flavours, top up with oil, and you will have luxurious delicacies for serving with salads, as appetizers, or as accompaniments. Sun-dried tomatoes also gain extra flavour when they are stored in oil. The concentrated tomato taste will soon brighten up the topping on a pizza, the filling in a quiche, or a sauce over pasta. There is an added advantage to be gained from this process. At the same time as being kept fresh by the oil, the preserved ingredients impart their delicious flavour to it. So, once they have been eaten, the flavoured oil can be used for cooking and in salad dressings.

Preserves in Oil

❦ Select the oil. The choice will depend upon how strong you want the flavour to be. Ordinary olive oil is much milder in flavour than extra-virgin olive oil. Corn oil or sunflower oil can be used if a neutrally flavoured oil is required. Nut oils are not a good choice, because they tend to go rancid much more quickly than other oils.

❦ Choose the ingredients. Vegetables can be preserved in oil, but use only those which can be processed in some way first to remove their excess water. The most usual methods are salting, such as with salt-cured olives and lemons, and drying, for vegetables like mushrooms and tomatoes. Mushrooms can also be cooked to extract their moisture. Feta and goat's cheeses take well to being preserved in olive oil and, after the cheese has been eaten, you can use the cheese-flavoured oil for cooking or in salad dressings. Once you have chosen which foods to preserve, marry them with some scented herb sprigs or pungent spices, to enhance their natural flavours.

❦ Pack the chosen ingredients and any seasonings into dry sterilized jars (see page 11), then pour in the oil to almost fill the jars and completely cover the ingredients.

❦ Leave for about 1 week in a cool dark place, so that the oil can preserve and flavour the ingredients, and the ingredients flavour the oil. The longer you leave them, the more potent the taste.

How to seal and store

Seal tightly in jars with non-corrosive, screw-top lids. Choose a cool dark place for storage, or store in the refrigerator. Foods preserved in oil will keep for up to 3 months. Any foods preserved in oil that contain garlic should always be stored in the refrigerator, and used within 1–2 weeks. You must make sure that the contents of all jars are always well covered with oil.

What can go wrong and why

If the oil goes rancid quickly, the jars have been stored in direct sunlight, which also destroys the vitamins in the oil. The oil may also go rancid if an opened bottle of oil is used for preserving the foods. Always break the seal on a brand new bottle of oil, so that you know it is fresh. If the contents of the jars start to go mouldy, it could be because they are not completely covered with oil.

Greek Olives in Oil

Olive oil, with its earthy aroma and flavour, is the most perfect complement to the grand and succulent Kalamata olive. It has a dual role in this recipe – to preserve the olives and absorb their flavour – so that once you have eaten the oil-soaked olives, you can use the scented oil for cooking.

Ingredients

675 g black Kalamata olives, in brine

1 tsp coriander seeds

1 large sprig of fresh rosemary

300–350 ml extra-virgin olive oil

Fills one 1 litre jar

1 Drain the olives into a sieve and rinse them well under cold running water. Drain again and dry the olives thoroughly on kitchen paper.

2 ◂Put the coriander seeds in a mortar and pestle and lightly crush them, being careful not to break them up completely. Alternatively, use the end of a rolling pin to break up the seeds.

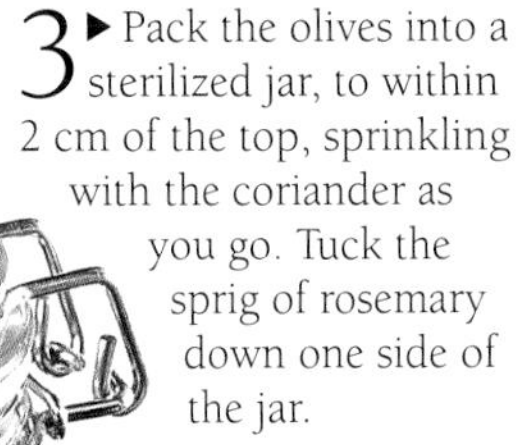

3 ▸ Pack the olives into a sterilized jar, to within 2 cm of the top, sprinkling with the coriander as you go. Tuck the sprig of rosemary down one side of the jar.

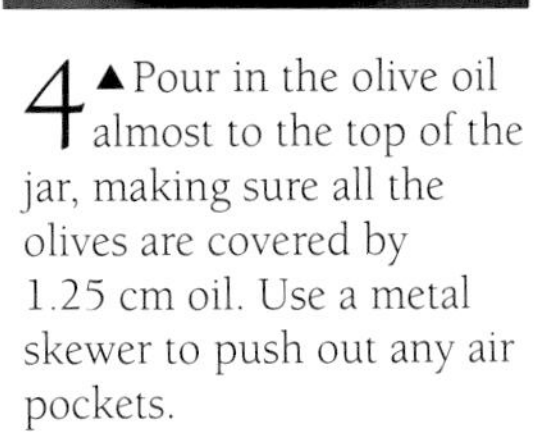

4 ▴Pour in the olive oil almost to the top of the jar, making sure all the olives are covered by 1.25 cm oil. Use a metal skewer to push out any air pockets.

5 Seal the jar and label. Keep in a cool dark place for at least 1 week before using, to allow the flavours to develop.

Red-Hot Lemon Slices in Olive Oil

Ingredients

6 lemons, total weight about 675 g

100 g sea salt

2 dried red chillies

2 dried bay leaves

1¼ tbsp paprika

1 tsp chilli powder

300 ml extra-virgin olive oil

1 Cut the lemons evenly into slices about 5 mm thick. Remove all the pips from the slices.

2 Put the lemon slices in a single layer in a shallow non-metallic dish. Sprinkle the slices with the salt, and leave to stand, covered, for 24 hours to draw out the excess moisture.

3 Drain the lemon slices, leaving the salt on the surface. Roughly chop the chillies and crumble the bay leaves into small pieces. Mix them in a small bowl with the paprika and chilli powder.

4 Carefully pack the salt-encrusted lemon slices in neat layers into sterilized jars, to within 2 cm of the tops, sprinkling the spice mixture evenly between the layers as you go.

5 Pour in the oil almost to the tops of the jars, making sure the lemons are covered by 1.25 cm. Seal the jars and label. Keep in a cool dark place for at least 1 week before using, to allow the flavours to develop.

Fills about two 480 ml jars

Sun-Dried Tomatoes in Olive Oil

Ingredients

50 g sun-dried tomatoes

2 bay leaves

200 ml extra-virgin olive oil

1 Put the sun-dried tomatoes into a bowl and cover with boiling water. Leave for 5 minutes, then drain well in a plastic sieve and pat dry on kitchen paper.

2 Pack the tomatoes into sterilized jars, to within 2 cm of the tops, tucking in the bay leaves as you go. Pour in the oil almost to the tops of the jars, making sure the tomatoes are covered by 1.25 cm.

3 Seal the jars and label. Keep in a cool dark place for at least 2 weeks before using, to allow the flavours to develop.

Fills about two 175 ml jars

Mushrooms in Olive Oil

Ingredients

675 g button mushrooms

480 ml extra-virgin olive oil

juice of 4 lemons

12 black peppercorns

4 peeled garlic cloves

2 bay leaves

1 Trim the mushroom stalks so that they are even with the caps. Put the mushrooms into a bowl. Add half of the oil, and stir in the remaining ingredients. Leave to stand for 3–4 hours.

2 Pour the mushroom mixture into a saucepan and bring to a boil. Simmer, stirring occasionally, for 15 minutes. Remove from the heat and leave to cool.

3 Drain the mushrooms in a plastic sieve and pat dry on kitchen paper. Pack them into a sterilized jar, to within 2 cm of the top. Pour in the remaining oil almost to the top of the jar, making sure the mushrooms are covered by 1.25 cm. Seal the jar and label. Refrigerate and use within 2 weeks.

Fills one 1 litre jar

Cook's Tip
Choose firm but moist mushrooms with no damp patches for this recipe, and wipe them clean with a damp tea-towel before using.

Goat Cheese with Herbs in Oil

INGREDIENTS

2 small round goat cheeses, each weighing about 3 oz

3 sprigs of fresh thyme

3 bay leaves

6 peppercorns

scant 1 cup extra-virgin olive oil

Leaving the cheese for 2–3 weeks ensures that the oil takes on the full flavor of the cheese and vice versa. Use the cheese in salads or on crisp French bread, and the oil for cooking.

1 Cut each goat cheese into quarters. Pack the pieces into a sterilized jar, layering the cheese with the herbs and peppercorns as you go.

2 Pour in the olive oil to cover the cheese by 1/2 inch.

3 Seal the jar and label. Keep in a cool dark place for 2–3 weeks before using to allow the flavors to develop.

Makes about 1 1/2 cups

Provençal Olives in Aromatic Oil

INGREDIENTS

1 1/2 lb green tanche olives, in brine

1 tsp fennel seeds

4 sprigs of fresh thyme

3 bay leaves

about 1 1/3 cups extra-virgin olive oil

1 Drain the olives in a sieve and rinse them well under cold running water to remove any brine. Dry them on paper towels.

2 Pack the olives into a sterilized jar, sprinkling with the fennel seeds and tucking in the thyme and bay leaves as you go.

3 Slowly pour in the olive oil to cover the olives by 1/2 inch. Use a metal skewer to push out any air pockets, taking care not to pierce the olives.

4 Seal the jar and label. Keep in a cool dark place for 1–2 weeks before using to allow the flavors to develop.

Makes about 1 quart

DID YOU KNOW?
Tanche olives are from Provence, in the south of France. If they are difficult to obtain, experiment with other green and black varieties.

Feta Cheese with Rosemary in Olive Oil

INGREDIENTS

6 oz feta cheese

4–6 sprigs of fresh rosemary

12 black peppercorns

12 black olives

scant 1 cup extra-virgin olive oil

1 Drain the cheese well and dry on paper towels. Cut the cheese into 3/4 inch cubes.

2 Pack the cheese cubes into sterilized jars, to within 3/4 inch of the tops, tucking in the rosemary, peppercorns, and olives as you go.

3 Pour in the olive oil to cover the cheese by 1/2 inch.

4 Seal the jars and label. Keep in a cool dark place for at least 1 week before using to allow the flavors to develop.

Makes about 2 cups

GREEK TREAT
Serve Feta Cheese with Rosemary in Olive Oil as a light lunch dish with warm continental breads.

Flavoured Vinegars

Cooks worldwide have always welcomed vinegar into their kitchens, and have created endless roles for it to play. Through the art of pickling, it is a great preserver of nature's bounty, as well as being a main ingredient for tenderizing tough meats, or making vinaigrette dressings for salads and crudités.

There are many basic types, from delicate Japanese rice vinegar through sharp wine vinegars to very robust malts, and exotic *aceto balsamico*. But add a variety of seasonings to flavour and colour vinegars, and they take on a whole new guise. Herbs, fruits, and spices – all of these can be steeped in vinegar. A dash of almost any sauce that has the blessing of a flavoured vinegar will banish blandness and add piquancy to soups, chowder, stews, and casseroles. Alternatively, use the vinegars combined with other seasonings to make marinades for meat, poultry, and seafood.

Making Flavoured Vinegars

❦ Sterilize the bottles (see page 11). Plain uncoloured glass bottles are the best choice to show off the colour and flavourings of the vinegar. Dry the bottles thoroughly before using.

❦ Choose a good-quality vinegar with an acetic acid content of 5% or more. Malt vinegar can be used to make pickling vinegar for pickles, but choose wine, sherry, or cider vinegar to make flavoured vinegars for use in salad dressings and for cooking.

❦ Select the flavourings. For herb vinegars, use fresh herbs, preferably picked before they flower. Use a single herb or several, mixing pungent and delicate ones together to balance the flavours. Fruit vinegars are usually made from soft fruits, such as raspberries, blueberries, or blackberries, or from citrus fruits. Other flavourings include garlic, dried chillies, and spices.

❦ Wash and dry the flavourings, as necessary. The herbs, fruits, and spices may be lightly crushed to release flavour.

❦ Insert the flavourings into the bottles, and add the vinegar almost to the tops.

❦ After sealing the bottles, the vinegar is left for 2 weeks before using, to allow the flavours to develop (unless the mixture is lightly cooked first, in which case the vinegar can be used immediately).

❦ Straining the vinegar after its initial storage ensures that it lasts longer and becomes clearer. You may need to strain the vinegar more than once to get a really clear liquid. After re-bottling, a few fresh ingredients can be added, both for decoration and identification. The bottles should then be re-sealed.

How to seal and store

Seal the bottles tightly with non-corrosive, screw-top lids, or new corks. Corks are only suitable for short-term storage, and they must be sterilized. After trimming the corks to fit the bottles, sterilize them in boiling water for a few minutes; this also softens them slightly. To seal, pound the corks into the necks of the bottles with a wooden mallet, leaving only 5 mm exposed cork. Flavoured vinegars should keep for about 12 months if the vinegar has been strained. If a large quantity of fresh herbs or garlic has been used, and the vinegar has not been strained, the shelf life will be shorter.

What can go wrong and why

The vinegar can ferment if it is kept too warm. A low storage temperature is also important for maintaining the flavour. Discard fermented vinegars or any that develop a questionable colour or flavour. Vinegar can evaporate from the bottle if it has not been sealed tightly enough.

BLUEBERRY-HERB VINEGAR

Colour and flavour pervade this distinguished vinegar, which combines plump blueberries, basil, and pink-tinted chive flowers. You can use one variety of basil, or a selection of sweet green and deep purple opal basil. For an elegant finishing touch, decant the vinegar into tall decorative bottles.

INGREDIENTS

1 large bunch of fresh basil

450 g blueberries

1 litre white wine vinegar

3 tsp chopped fresh chives

fresh chive flowers and blueberries, to finish

Makes about 900 ml

1 ▲Strip the basil leaves from their stalks. Tear the leaves into small pieces. Discard the stalks.

2 ▲Put the blueberries and a little vinegar into a non-metallic bowl. Crush the berries with the back of a wooden spoon to release their juices.

Stir lightly to mix colours and flavours

3 ◀ Stir in the remaining vinegar, the basil, and the chopped chives. Pour the mixture into a large sterilized jar. Seal and shake well. Keep in a cool dark place for 4 weeks before using, to allow the flavours to develop. Shake the jar from time to time.

4 ▶Line a funnel with a double layer of muslin. Strain the flavoured vinegar through it into sterilized bottles, to within 3 mm of the tops. Add a fresh chive flower and a few blueberries to each bottle to give an attractive finish. Seal and label. The vinegar is now ready to use.

HERBS OF PROVENCE VINEGAR

INGREDIENTS

4 large sprigs of fresh rosemary

4 large sprigs of fresh tarragon

4 sprigs of fresh thyme

6 fresh bay leaves

2 pinches of fennel seeds

1 litre red wine vinegar

Aromatic flavours from the south of France are combined here to give red wine vinegar a regional taste.

1 Lightly bruise the rosemary, tarragon, thyme, and bay leaves to release their flavours. Put all the herbs and the fennel seeds into a sterilized jar, then pour in the vinegar.

2 Seal the jar and shake well. Keep in a cool dark place for 2–3 weeks before using, to allow the flavours to develop. Shake the jar from time to time.

3 Line a funnel with a double layer of muslin. Strain the flavoured vinegar through it into sterilized bottles, to within 3 mm of the tops. Seal the bottles and label. The vinegar is now ready to use.

Makes about 1 litre

DID YOU KNOW?
The word vinegar comes from the French *vinaigre,* or sour wine. Half of the wine vinegar produced in France comes from the city of Orléans in the Loire valley.

SPICED BLACKBERRY VINEGAR

INGREDIENTS

1 kg blackberries

2 cinnamon sticks

2 tsp allspice berries

2 tsp whole cloves

600 ml white wine vinegar

450 g sugar

a few fresh blackberries, to finish

1 Pick over the blackberries, removing any mouldy or damaged parts. Break the cinnamon sticks into pieces. Put all the spices on a square of muslin and tie up tightly with a piece of string.

2 Put the vinegar and sugar into a large saucepan. Stir over a low heat, until the sugar has completely dissolved. Add the spice bag. Bring to a boil, lower the heat, and simmer for 5 minutes.

3 Add the blackberries and simmer for a further 10 minutes. Remove from the heat and leave the blackberry mixture to cool completely. Discard the spice bag.

4 Line a funnel with a double layer of muslin. Strain the flavoured vinegar through it into sterilized bottles, to within 3 mm of the tops. Add fresh blackberries to each bottle. Seal the bottles and label. The vinegar is now ready to use.

Makes about 1 litre

ORANGE-SCENTED VINEGAR

INGREDIENTS

4 small oranges

1 litre white wine vinegar

freshly pared orange zest, to finish

1 Thinly pare the zest from the oranges with a vegetable peeler, without taking off any pith. Cut the oranges in half and squeeze out the juice. Put the orange zest and juice into a sterilized jar, then pour in the vinegar.

2 Seal the jar and shake well. Keep in a cool dark place for at least 4 weeks before using, to allow the flavours to develop. Shake the jar from time to time.

3 Line a funnel with a double layer of muslin. Strain the flavoured vinegar through it into sterilized bottles, to within 3 mm of the tops.

4 Add 1–2 strips of freshly pared orange zest to each bottle. Seal the bottles and label. The vinegar is now ready to use.

Makes about 1 litre

GARLIC VINEGAR

INGREDIENTS

16 garlic cloves

salt

1 litre white or red wine vinegar

GARLIC STICK
For an unusual presentation, thread 2–3 peeled garlic cloves on to a wooden skewer and drop into the bottle of Garlic Vinegar.

The flat side of a chef's knife can be used to crush garlic cloves. The granular texture of a little salt helps when crushing the cloves.

1 Peel the garlic cloves and lightly crush them with a little salt. Put the garlic into a large sterilized jar, then pour in the vinegar.

2 Seal the jar and shake well. Keep in a cool dark place for 2–3 weeks before using, to allow the flavours to develop. Shake the jar from time to time.

3 Line a funnel with a double layer of muslin. Strain the vinegar through it into sterilized bottles, to within 3 mm of the tops. Seal the bottles and label. The vinegar is now ready to use.

Makes about 1 litre

ROSEMARY AND ALLSPICE VINEGAR

INGREDIENTS

3–4 sprigs of fresh rosemary

1½ tsp allspice berries

480 ml red or white wine vinegar

1 Strip the rosemary leaves from their stalks. Lightly bruise the leaves to release their flavour. Put the allspice berries in a mortar and pestle and lightly crush them, being careful not to break them up completely. Alternatively, use the end of a rolling pin to crush them.

2 Put the rosemary and allspice into a sterilized jar, then pour in the vinegar. Seal the jar and shake well. Keep in a cool dark place for 4 weeks before using, to allow the flavours to develop. Shake the jar from time to time.

3 Line a funnel with a double layer of muslin. Strain the vinegar through it into a sterilized bottle, to within 3 mm of the top. Seal the bottle and label. The vinegar is now ready to use.

Makes about 480 ml

SPICED PICKLING VINEGAR

INGREDIENTS

2 cinnamon sticks

2 blades of mace

1 tbsp whole cloves

1 tbsp allspice berries

3 tsp black peppercorns

1.15 litres cider vinegar

This vinegar is used to flavour vegetable pickles, chutneys, and even fruit pickles. It is also delicious when used with extra-virgin olive oil to make a vinaigrette dressing for salads.

1 Divide the cinnamon sticks, blades of mace, whole cloves, allspice berries, and black peppercorns equally between 2 sterilized bottles. Slowly pour in the vinegar, to within 3 mm of the tops.

2 Seal the bottles and shake well. Label the bottles and keep in a cool dark place for 2 months before using, to allow the flavours to develop. Shake the bottles from time to time. The vinegar is now ready to use.

Makes about 1.15 litres

COOK'S TIP You can use the vinegar in 24 hours, rather than leaving it to mature, by heating it to just below boiling point, then pouring it over the spices. It has even more flavour if left for a week.

Tarragon Vinegar

INGREDIENTS

1 bunch of fresh tarragon, weighing about 25 g

480 ml white wine vinegar

Fresh chervil can be substituted for tarragon in this recipe.

1 Lightly bruise the tarragon to release its flavour. Put the tarragon into a sterilized jar. Pour in the vinegar.

2 Seal the jar and shake well. Keep in a cool dark place for 2–3 weeks before using, to allow the flavours to develop. Shake the jar from time to time.

3 Line a funnel with a double layer of muslin. Strain the vinegar through it into a sterilized bottle, to within 3 mm of the top. Seal the bottle and label. The vinegar is now ready to use.

Makes about 480 ml

Bouquet of Herbs Vinegar

INGREDIENTS

4 large sprigs each of fresh parsley, rosemary, tarragon, and thyme

12 black peppercorns

4 celery sticks

4–6 shallots, depending on size

1 litre white wine vinegar

1 Lightly bruise the herbs to release their flavour. Lightly crush the peppercorns in a mortar and pestle, being careful not to break them up completely. Alternatively, use the end of a rolling pin to crush them. Thinly slice the celery and shallots.

2 Put all the flavourings into a large sterilized jar, then pour in the vinegar. Seal the jar and shake well. Keep in a cool dark place for 2–3 weeks before using, to allow the flavours to develop. Shake the jar from time to time.

3 Line a funnel with a double layer of muslin. Strain the flavoured vinegar through it into sterilized bottles, to within 3 mm of the tops. Seal and label the bottles. The vinegar is now ready to use.

Makes about 1 litre

Rosy Raspberry Vinegar

INGREDIENTS

450 g raspberries

480 ml white wine vinegar

50 g sugar

a few fresh raspberries, to finish

This is a concentrated vinegar that should be used sparingly. The addition of raspberries yields a delicious flavour and colour.

1 Pick over the raspberries and remove any mouldy or damaged parts. Put the berries and a little vinegar into a non-metallic bowl. Crush the berries with the back of a wooden spoon, to release their juices. Stir in the remaining vinegar.

2 Pour the mixture into a sterilized jar. Seal the jar and shake well. Keep in a cool dark place for 2 weeks before using, to allow the flavours to develop. Shake the jar from time to time.

3 Line a funnel with a double layer of muslin. Strain the flavoured vinegar through it into a small saucepan. Stir in the sugar and simmer over a low heat for 10 minutes. Allow to cool.

4 Pour the sweetened vinegar into a sterilized bottle, to within 3 mm of the top. Add a few fresh raspberries to the bottle. Seal the bottle and label. The vinegar is now ready to use.

Makes about 300 ml

Herbs & Spices

Herb & Spice Blends

Herbs and spices add flavour and variety to foods that would be very dull without them. Herbs have many virtues: they contribute vitamins to our daily fare, they originate from all over the world but seldom fail to adapt to different soils and climates, and they often have a natural culinary affinity for each other. The classic combination of herbs found in most kitchens – a few stalks of parsley, a sprig of thyme, and a bay leaf, all tied together with a piece of thread or string – is known by French chefs as a *bouquet garni*. If the herbs are dried they are wrapped in a square of muslin and tied with string. Another popular combination in France is *herbes de Provence*, a mixture of the herbs that grow in the hills of this southern region of France, so it is especially good in dishes from the Mediterranean.

Spices, close kitchen companions of herbs, are more robustly flavoured and more exotic because many are from the tropics. Although very different in nature, they combine successfully with sugar. Spices are used in cakes and biscuits, to flavour wine punches, and particularly in Asian cooking. They are an essential component of curries, and they season cuisines worldwide. Herb and spice blends give every cook a veritable battalion of kitchen condiments for livening up their dishes.

Making Herb & Spice Blends

❦ Select the ingredients. Buy herbs and spices at a shop where you know there is a high turnover, so that they are really fresh. If possible, grow and dry your own herbs.

❦ Prepare the herbs or spices. Crumble dried herbs between your fingers or crush them with a rolling pin. Large amounts are easier to crush if put in a plastic bag first. Process dried spices in a mortar and pestle or in an electric grinder. Alternatively, the end of a rolling pin can be used for crushing spices. This treatment also helps to prevent peppercorns, allspice berries, and coriander seeds from getting stuck in the blades of the grinder. Occasionally, the mixed herbs and spices are dry-fried before they are ground; this helps to extract the full flavour of the ingredients.

❦ Mix the herbs or spices in a bowl. Do not stir too vigorously or the powders will make you sneeze. Spoon the mixture into small sterilized jars (see page 11), packing it down well by tapping the bottom of the jar on the work surface. A hand-made funnel placed in the top of the jar makes the job of spooning in spices easier.

How to seal and store

Seal the jars tightly with screw-top lids. These are the best choice for keeping herb and spice blends fresh. Corks can also be used, trimmed to fit the necks of the jars, and they look attractive, but the spices will lose their flavour more quickly. Keep the jars in a cool dark place, for up to 4–6 months for herb and spice blends, and up to 1 year for flavoured sugars.

What can go wrong and why

If the herb and spice blends lose their colour and flavour, they have been stored for too long or incorrectly. Unless they are kept in dry conditions in completely airtight containers, the flavoured salts and sugars will become compact and will not pour freely.

Aromatic Garam Masala

The blending of spices is an integral part of all Indian cooking. In northern India, the most important mixture of spices is known as garam masala: "masala" literally means a "blend". Each cook creates his or her own blend, making it fiercely fiery or subtly aromatic, from a selection of two or three spices to sometimes over a dozen. This mild mix, dominated by the scent of cardamom, is ideal for meat curries.

Ingredients

20 green cardamom pods
3 cinnamon sticks
4 dried bay leaves
1 1/2 tbsp black peppercorns
1 tbsp cumin seeds
2 tsp whole cloves
2 tsp freshly grated nutmeg
Makes about 40 g

1 Split open the cardamom pods with a small sharp knife and remove the dark brown seeds. Discard the pods and crush the seeds in a mortar and pestle. Alternatively, use the end of a rolling pin to crush them.

2 ◀ With your fingers, break the cinnamon sticks into fairly small lengths. Crumble each dried bay leaf into several small pieces.

Stir and shake the pan frequently when frying the spices, to keep them on the move and prevent them from burning.

3 ▶ Put all the spices, except the nutmeg, into a heavy frying pan. Dry-fry them over a medium heat for 2–3 minutes. Remove the pan from the heat and put the spices into a small bowl. Allow to cool, and stir in the grated nutmeg.

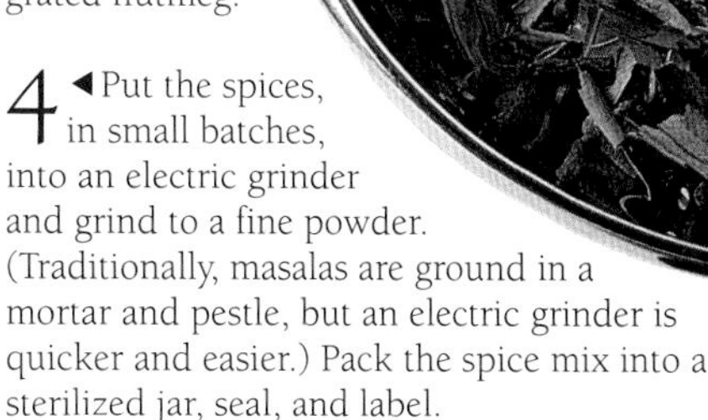

4 ◀ Put the spices, in small batches, into an electric grinder and grind to a fine powder. (Traditionally, masalas are ground in a mortar and pestle, but an electric grinder is quicker and easier.) Pack the spice mix into a sterilized jar, seal, and label.

KASHMIRI MASALA

INGREDIENTS

8 green cardamom pods

1 cinnamon stick

1 1/2 tbsp cumin seeds

3 tsp black peppercorns

2 tsp whole cloves

2 tsp caraway seeds

1 tsp freshly grated nutmeg

The fragrance of cardamom permeates Kashmiri Masala, a blend of spices from the northernmost valleys of India. Green cardamoms have the most delicate flavour, while the brown have an unpleasant taste and should not be used. This masala is good for chicken and lamb curries.

1. Remove the seeds from the cardamom pods and crush them in a mortar and pestle. Alternatively, use the end of a rolling pin. Break the cinnamon stick into several pieces.
2. Put all the spices, except the nutmeg, into a heavy frying pan and dry-fry over a medium heat for 2–3 minutes. Allow to cool.
3. Put the whole spices, and the nutmeg, into an electric grinder and grind to a fine powder. Alternatively, use a mortar and pestle. Pack into a sterilized jar, seal, and label.

Makes about 30 g

CHAAT MASALA

INGREDIENTS

6 dried red chillies

90 ml cumin seeds

90 ml coriander seeds

1 tbsp black peppercorns

"Dry-frying" whole spices in a frying pan to a dark brown colour before grinding, as in this masala recipe, extracts their full flavour and heightens their aroma.

1. Break the chillies into pieces. Dry-fry all the spices in a heavy frying pan over a medium heat until the seeds start to pop and colour. This takes 2–3 minutes. Remove the spices from the heat so that they do not burn. Allow to cool.
2. Put all the spices into an electric grinder and grind to a fine powder. Alternatively, use a mortar and pestle. Pack into a sterilized jar, seal, and label.

Makes about 65 g

PICKLING SPICE

INGREDIENTS

4 blades of mace

2 cinnamon sticks

2 small dried red chillies

1 1/2 tbsp ground ginger

1 1/2 tbsp allspice berries

1 1/2 tbsp whole cloves

1 1/2 tbsp coriander seeds

1 1/2 tbsp mustard seeds

1 1/2 tbsp black peppercorns

In European cooking, the idea of mixing different spices together to make a blend has fallen from favour over the years, with cooks preferring to use spices individually. One of the mixes that has retained its popularity is Pickling Spice, used to liven up the flavour of chutneys, pickles, and vinegars.

1. Break the mace blades, cinnamon sticks, and chillies into pieces. Put all the spices into a bowl and stir. Pack into a sterilized jar, seal, and label.
2. To use, put the measured spices on a square of muslin and tie up tightly with a long piece of string. Add to the recipe when specified. Remove the bag after pickling.

Makes about 40 g

VARIATION Proportions and types of spices for this traditional English recipe can vary. Fennel seeds can be added, and a pinch of freshly grated nutmeg can replace the mace.

Five-Spice Powder

INGREDIENTS

2 cassia sticks

6 star anise

3 tsp whole cloves

3 tsp fennel seeds

3 tsp anise seeds or Szechuan pepper or whole black peppercorns

When blended together according to an ancient formula, these five spices create a harmonious mix of bitter, sweet, sour, and salty flavours. The pungent taste of Five-Spice Powder permeates many Chinese and Vietnamese roast meat and poultry dishes.

1 Break the cassia sticks into several pieces.

2 Put all of the spices into an electric grinder and grind to a fine powder. Alternatively, use a mortar and pestle.

3 Pack into a sterilized jar, seal, and label.

Makes about 25 g

VARIATION Cinnamon can be used instead of cassia. Use slightly more because the taste is milder.

Quatre Epices

INGREDIENTS

2 tbsp white peppercorns

1 tsp whole cloves

1 tbsp freshly grated nutmeg

3 tsp ground ginger

Like the Pickling Spice mix, this French blend of four spices can vary in its composition. Allspice and cinnamon can be substituted for any of the spices in this recipe.

1 Put the peppercorns and cloves in an electric grinder and grind to a fine powder. Alternatively, use a mortar and pestle.

2 Put into a bowl and mix with the freshly grated nutmeg and the ground ginger.

3 Pack into a sterilized jar, seal, and label.

Makes about 40 g

Dill Pickle Spices

INGREDIENTS

2 small red chillies

4 dried bay leaves

3 tsp ground ginger

2 tbsp mustard seeds

3 tsp dill seeds

3 tsp coriander seeds

2 tsp black peppercorns

1 tsp allspice berries

1 tsp whole cloves

1/2 tsp fennel seeds

The dainty dill seed, with a flavour reminiscent of caraway, is the special ingredient for Dill Pickle Spices. Use this blend for pickled cucumber or dill pickles, pickled gherkins, or in flavoured vinegars.

1 Break the chillies and bay leaves into pieces. Put all spices in a bowl, and mix well.

2 Pack into a sterilized jar, seal, and label.

Makes about 50 g

Left to right, *Quatre Epices, Kashmiri Masala, Dill Pickle Spices*

All-American Barbecue Mix

INGREDIENTS

$1^1/_2$ tbsp each of dried parsley and chives

3 tsp each of dried mint, thyme, and tarragon

$1^1/_2$ tbsp freshly ground black pepper

1 tsp paprika

1 Put all the ingredients into a bowl and mix together well, making sure to lightly crush the dried herbs into several pieces.

2 Pack the mixture into a sterilized jar. Seal the jar and label. The barbecue mix is now ready to use.

Makes about 25 g

VARIATIONS Different dried herbs and spices, such as rosemary, cumin, and chilli, can be used to make this mix. Other flavourings like honey and mustard can also be added to the mix, which is sprinkled directly on to meat and poultry while barbecuing.

Seven Seas Spice Mix

INGREDIENTS

15 green cardamom pods

1 cinnamon stick

$1^1/_2$ tbsp coriander seeds

2 dried red chillies

3 tsp cumin seeds

2 tsp each of celery seeds and whole cloves

This fragrant mix is an excellent flavouring for Indonesian, Malaysian, and Korean dishes.

1 Remove the seeds from the cardamom pods and crush them in a mortar and pestle. Alternatively, use the end of a rolling pin. Break the cinnamon stick into several pieces.

2 Put all the spices into an electric grinder and grind to a fine powder. Alternatively, use a mortar and pestle. Pack spice mixture into a sterilized jar. Seal the jar and label. The spice mix is now ready to use.

Makes about 40 g

Tarragon Salt

INGREDIENTS

1 bunch of fresh tarragon, weighing about 50 g

115 g sea salt

1 Set the oven to its lowest temperature. Strip the tarragon leaves from the stalks. Discard the stalks. Coarsely chop the leaves and mix with the salt in a blender until the leaves are finely chopped.

2 Spread the leaves out on a baking sheet covered with aluminium foil. Put in the oven with the door ajar and leave for $1^1/_2$ hours, or until crisp and dry. Allow to cool.

3 Pack the salt into sterilized jars. Seal the jars and label. The tarragon salt is now ready to use.

Makes about 150 g

Seasoning Salt

INGREDIENTS

75 g sea salt

$1^1/_2$ tsp each of ground celery seeds, white pepper, cumin, and paprika or cayenne pepper

This is an important condiment for marinating meats.

1 Put all the ingredients into a bowl and mix together well.

2 Pack into sterilized jars. Seal the jars and label. The seasoning salt is now ready to use.

Makes about 90 g

ITALIAN SEASONING

INGREDIENTS

8–12 dried bay leaves

2 tbsp each of dried oregano, thyme, and sage

2 tbsp each of freshly ground pepper and paprika

1 Crush the bay leaves with a rolling pin until they are broken into quite fine pieces. Alternatively, use a mortar and pestle. Put them with all the remaining ingredients into a bowl and mix together well.

2 Pack into a sterilized jar. Seal the jar and label. The Italian seasoning is now ready to use.

Makes about 40 g

HERBES DE PROVENCE

INGREDIENTS

3 tbsp each of dried oregano, savory, thyme, marjoram, and rosemary

In Provence, dried herbes de Provence *are sold in little terracotta pots topped with the local patterned cloth, or in brightly coloured bags of the same material. By a simple blending of five herbs, you can make your own blend, for adding authenticity to any Provençal dish.*

1 Put all the herbs into a bowl and mix together well.

2 Pack into a sterilized jar. Seal the jar and label. The herbs are now ready to use.

Makes about 20 g

BOUQUET GARNI

INGREDIENTS

12 dried bay leaves

3 tbsp dried thyme

3 tbsp dried parsley

1 1/2 tbsp dried celery leaves

This French blend is usually tied together in a muslin square when dried ingredients are used. The bundle should always contain a bay leaf, thyme, and parsley. Sometimes other herbs, like chervil, savory, or tarragon, are added. Celery leaves can also be used.

1 Crush the bay leaves with a rolling pin until they are broken into quite fine pieces. Alternatively, use a mortar and pestle. Put them with the remaining herbs into a bowl and mix together.

2 Cut a double thickness of muslin into 7.5 cm squares. Put 3 tsp mixture on each square of muslin and tie up into bags with string. Pack the bags into sterilized jars. Seal the jars and label. The bouquets garnis are now ready to use.

Makes about 15 g, enough for 12 muslin bags

ENGLISH MIXED HERBS

INGREDIENTS

90 ml each of dried parsley, chives, thyme, and tarragon

This is the English version of herbes de Provence. *It is a wonderful complement for lamb, pork, or stuffing. Rosemary, sage, and marjoram can be used in addition, or as substitutes.*

1 Put all the herbs into a bowl and mix together well.

2 Pack into a sterilized jar. Seal the jar and label. The mixed herbs are now ready to use.

Makes about 25 g

Mixed Spices

INGREDIENTS

2 cinnamon sticks

$1^1/_2$ tbsp coriander seeds

2 tsp each of allspice berries and whole cloves

1 tbsp ground ginger

1 tsp freshly grated nutmeg

This spice blend has been popular in English cooking since the sixteenth century. Few traditional mixtures have survived as long. Use it in puddings, cakes, and biscuits.

1 Break the cinnamon sticks into several pieces. Put with the coriander, allspice, and cloves into an electric grinder and grind to a fine powder. Alternatively, use a mortar and pestle.

2 Put the spice mixture into a bowl and stir in the ginger and nutmeg. Pack into a sterilized jar. Seal the jar and label. The spices are now ready to use.

Makes about 25 g

Vanilla Sugar

INGREDIENTS

1 kg caster sugar

4 vanilla pods

As this vanilla sugar is used up, in custards, cakes, and milk puddings, keep the jar topped up with fresh caster sugar. It will continue to absorb the vanilla flavour from the pods.

1 Divide the sugar between 2 sterilized jars. Press 2 vanilla pods down into the sugar in each jar.

2 Seal the jars and label. Keep in a cool dark place for 1 week before using, to allow the flavours to develop.

Fills two 500 ml jars

Cinnamon Sugar

INGREDIENTS

1 kg caster sugar

100 g cinnamon sticks

1 Divide the sugar between 2 sterilized jars. Divide the whole cinnamon sticks evenly between the jars and press the sticks down into the sugar.

2 Seal the jars and label. Keep in a cool dark place for 1 week before using, to allow the flavours to develop.

Fills two 500 ml jars

Orange or Lemon Sugar

INGREDIENTS

6 oranges or 8 lemons

1 kg caster sugar

1 Set the oven to its lowest temperature. Peel the zest from the fruit with a vegetable peeler. You will need about 100 g zest. Spread the zest out on a baking sheet covered with aluminium foil. Put in the oven and leave for about 3 hours, or until dried out. Let the zest cool.

2 Divide the sugar between 2 sterilized jars, layering it with the dried zest. Seal the jars and label. Keep in a cool dark place for 1 week before using, to allow the flavours to develop.

Fills two 500 ml jars

Spice Mix for Mulled Beer

Ingredients

6 cinnamon sticks

12 pieces of dried lemon zest (see box, page 110)

6 whole nutmegs, cut in half

36 whole cloves

To make mulled beer: use 1 bag of mix for each 600 ml beer.

1 Cut a double thickness of muslin into 12 pieces, each about 5 cm by 7.5 cm.

2 Break each cinnamon stick into 2 pieces. Put 1 piece each of cinnamon and lemon zest with half a nutmeg and 3 cloves on each piece of muslin. Gather the muslin into small bundles and tie up tightly into bags with string. Pack the bags into sterilized jars. Seal the jars and label. The spice bags are now ready to use.

Makes about 50 g, enough for 12 muslin bags

Spice Mix for Glühwein

Ingredients

24 small pieces of dried root ginger

90 ml allspice berries

24 whole cloves

This German spice blend is traditionally used to make Glühwein.

1 Put all of the spices into a bowl and mix together well.

2 Pack into sterilized jars. Seal the jars and label. The spice mix is now ready to use.

Makes about 50 g

Winter Warmer
Glühwein is the perfect beverage when the weather is cold.

To Make *Glühwein* Pour 2 bottles of red wine into a saucepan, and add 2 tbsp sugar. Simmer over a low heat, stirring with a wooden spoon, until the sugar has completely dissolved. Do not allow the mixture to boil. Add 2 tbsp *Glühwein* spices, 1 sliced orange and 1 sliced lemon. Heat to just below boiling point. Taste and add more sugar, if necessary. Leave the *Glühwein* to stand for 1 hour, then reheat and serve. Makes 6–8 servings.

Orange Spice Mix

Ingredients

12 cinnamon sticks

24 whole cloves

2 tbsp allspice berries

12 pieces of dried orange zest (see box, page 110)

1 Break each cinnamon stick into 2 pieces. Put the broken cinnamon sticks with the remaining ingredients into a small bowl and mix together well.

2 Pack the mixture into sterilized jars. Seal the jars and label. The spice mix is now ready to use.

Makes about 75 g

Cook's Tip When infusing these spices in punches, allow 1½ tbsp mixture for each 600 ml liquid. Tie up the spices into a muslin bag, so the spices do not have to be strained out.

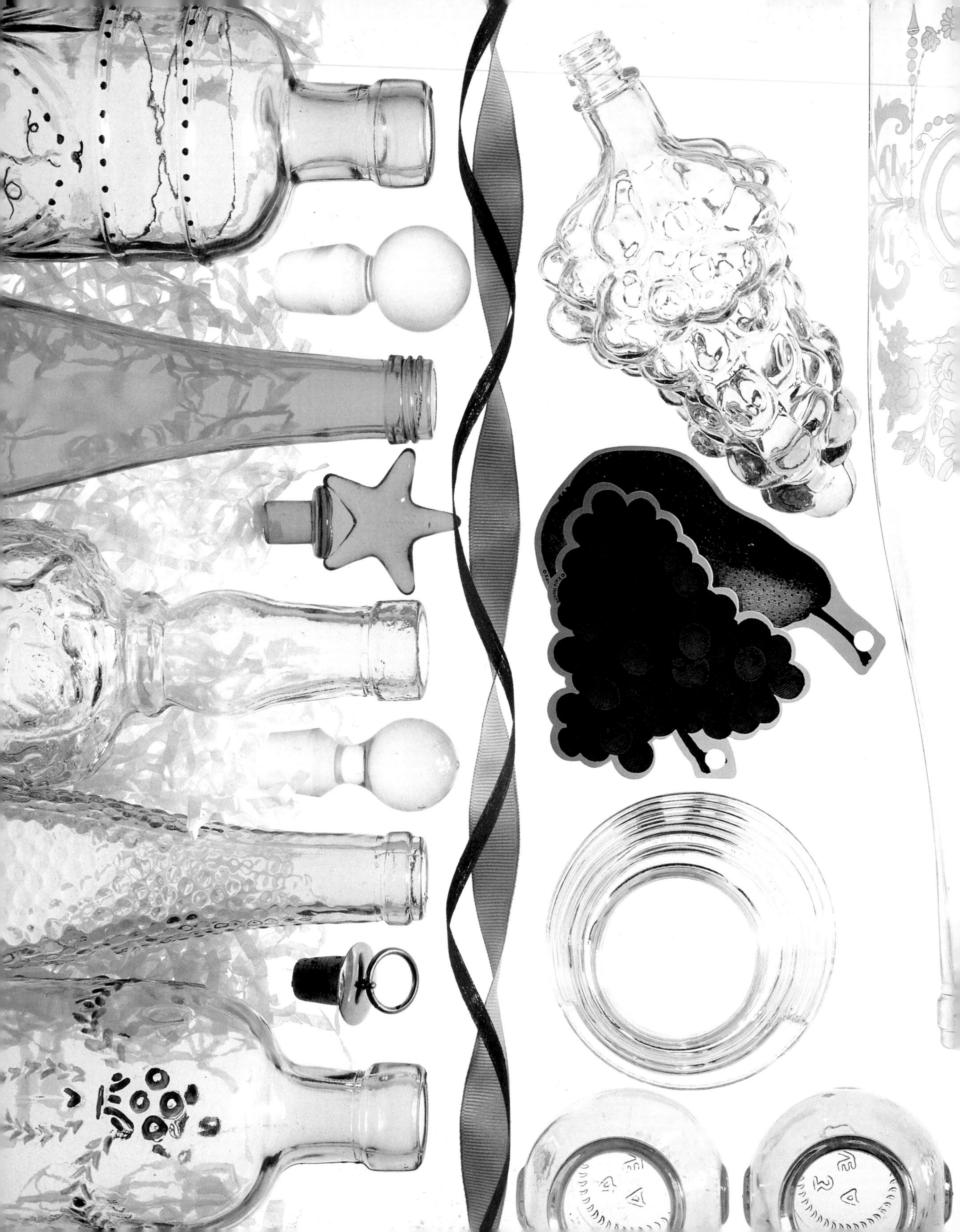

Finishing Touches

Decorating Bottles

Glass bottles can be transformed into rustic, romantic, vibrant, or psychedelic creations by a little paintwork, stencilling, or spraying. All you need are metallic spray paints, waterproof felt-tip pens, sable-hair brushes, and specially formulated glass paints, which are all available in good craft shops. Use clean dry bottles so that the paint will adhere properly. Handmade paper collars are an original finishing touch.

Above left, *hand-painted bottle;* **above right**, *negative-stencilled bottle*

Positive Stencilling

Fix a stencil on to the bottle and spray with metallic spray paint, to simulate an etched or antique glass effect. Be sure to cover all other surfaces of the bottle with adhesive tape, protecting surfaces with newspaper, and your hands with rubber gloves. When the paint is dry, remove the tape and stencil. Any blurred edges can be cleaned up with a razor blade or with a little nail polish remover.

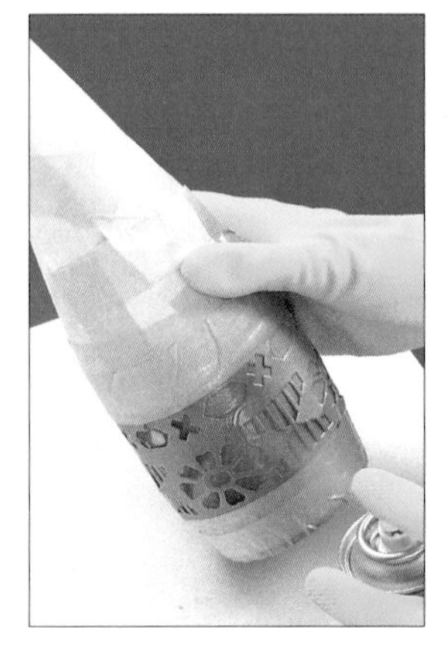

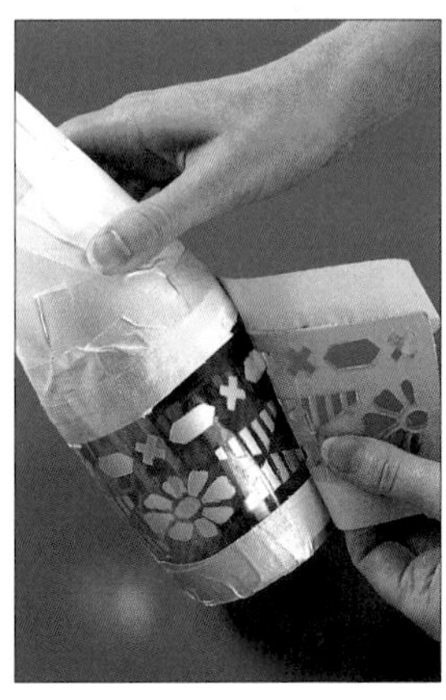

Negative Stencilling

Stick self-adhesive motifs, or shapes made from masking tape, all around the bottle to create a pattern. Stick the edges down well, or paint will seep underneath. Protect the top of the bottle with tape. Spray the entire bottle with metallic spray paint, leave for 30 minutes, and spray with a second coat. When the paint is dry, carefully remove the motifs to reveal the pattern on the bottle.

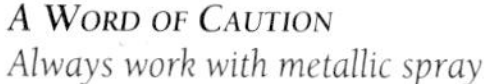

A Word of Caution

Always work with metallic spray paints in a well-ventilated area, preferably outside. Protect all surrounding surfaces and wear rubber gloves.

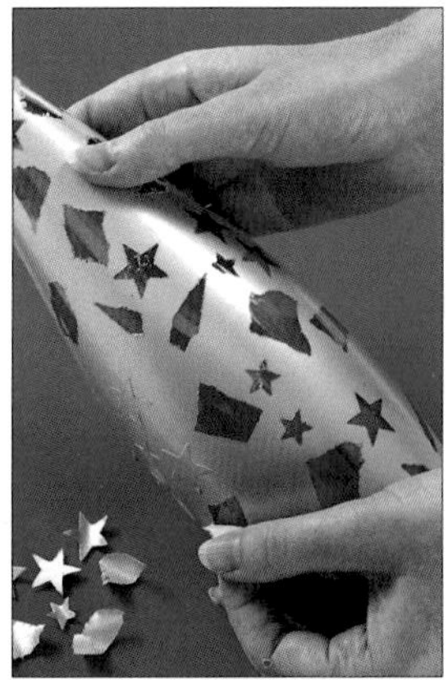

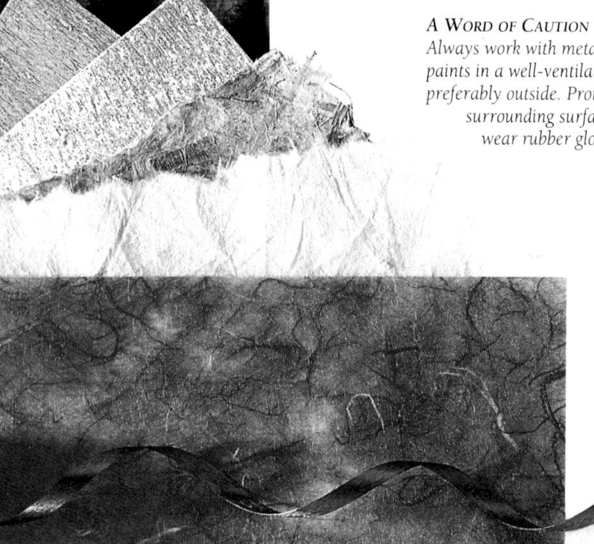

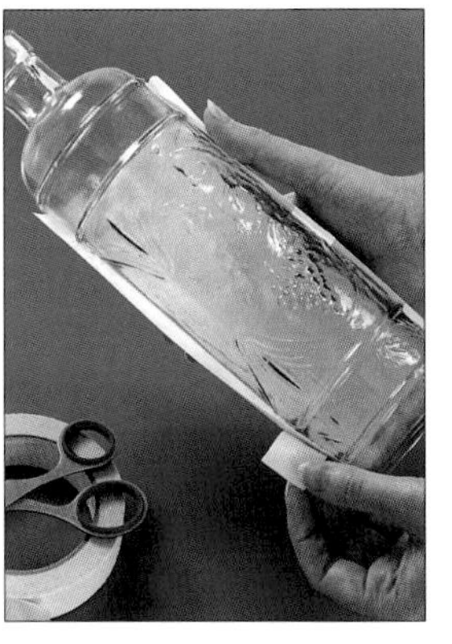

PAINTING BOTTLES

Use formulated glass paints or waterproof felt-tip pens to decorate bottles. Try simple patterns before graduating to more complicated designs. Apply the paints in one stroke because they dry quickly. Some thinners that delay drying times are available, but use sparingly. Highlight the design on the bottle by attaching white card to the other side, removing the card for the finishing touches. Delicate hand-painted surfaces can be protected with a coat of clear craft varnish.

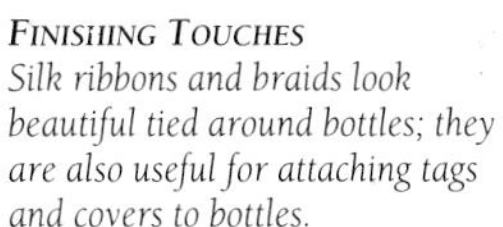

FINISHING TOUCHES
Silk ribbons and braids look beautiful tied around bottles; they are also useful for attaching tags and covers to bottles.

MAKING COVERS AND COLLARS

Covers for bottles can be made with material or metallic crêpe paper. Cut a circle from the material or paper and tie on to the top of the bottle with a large ribbon or cord. Collars, made from card, are more difficult: draw a template in the shape shown below, on to the reverse of some coloured card. Cut out the template. Lightly score along the dotted lines and bend the card. Glue or tape the angled end to the inside of the opposite end, to fasten and slip over the bottle neck.

PRETTY AS A PICTURE
Spray-painted and hand-painted bottles look artistic when finished with a collar or decorated with leaves and braiding.

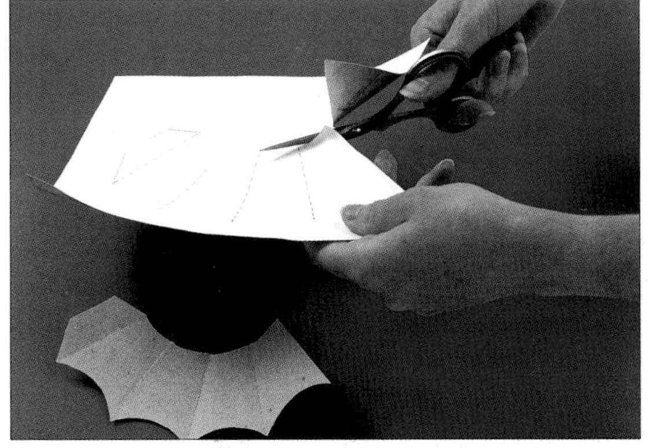

Decorating Jars

Gift wrapping does not need to cost a fortune if you hunt in the right places. Liven up a humble jar with a spray of découpage cherries, a leaf-wrapped lid, or a bright strip of Petersham ribbon sealed with wax. Alternatively, a cotton bandeau ribbon or a silk cord can be tied around the jar to make it look extra special.

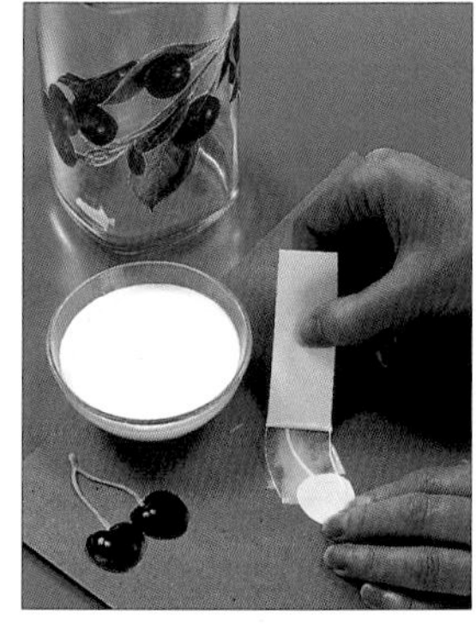

Découpage

Decorating surfaces with paper cut-outs is known as découpage. Cut out fruit or vegetable pictures from magazines, keeping the edges neat (a sharp scalpel is best for this task). Coat the back of the cut-out with paper adhesive, and position decoratively on the jar or lid.

Picturesque Preserves
Paper cut-outs can turn preserves into fancy gifts. They look beautiful and also indicate what is inside.

Covering Lids with Leaves

Wash and dry large leaves, such as Virginia creeper, sycamore, or maple. Paint the dry leaves on both sides with glycerine (available from chemists). Arrange them on the top of the lid, wrap with cling film, and secure with a rubber band. Weight the lid and leave in the refrigerator overnight. Gently unwrap, and wipe off any excess glycerine. Secure the leaf on the lid with a piece of raffia.

Making a Wax Seal
Wrap ribbon around the jar and hold temporarily in place with masking tape. Use a taper to light a stick of sealing wax and hold it about 2.5 cm away from the point to be sealed. Let the wax drip steadily until a 2.5 cm wide blob has formed. If the flame goes out, re-light the wax and continue. Before the wax has time to cool, press a decorative metal seal down firmly on the melted wax, and hold it there for a few seconds until a good impression has been left. Remove the masking tape. *Note: it is a good idea to practise making your seals beforehand with spare ribbon on empty jars.*

Sealing with Wax
A variety of decorative metal seals and wax sticks are available from specialist stationery shops.

***Far right**, wax seals add a professional touch to gifts; **right**, leaf-wrapped lids are an unusual but natural covering for jars*

PRESENTATION IDEAS

A beautifully presented preserve is a pleasure both to give and receive. Here are some imaginative and innovative ways to make your gifts look elegant and stylish.

ALL WRAPPED UP
Make your own gift bags with wallpaper and string, and place preserves inside in a straw "nest". Or, cover with unbleached linen and tie with raffia.

IN THE BAG
Use clear paper adhesive to stick shells and starfish on to bottle bags. Rest bottles in shredded tissue.

HAT TRICK
Wrap hat netting around a bottle and secure with florist's wire. Decorate with braid and ribbon.

A TOUCH OF SPICE
Spice bags are made of muslin tied with ribbons or strings for gift-giving. Place several bags in a glass and wrap with patterned cellophane. Bundle up with a beautiful bow.

BOX CLEVER
Cover boxes with designer paper, pleating the edges. Use doilies and tissue to create a lavish effect.

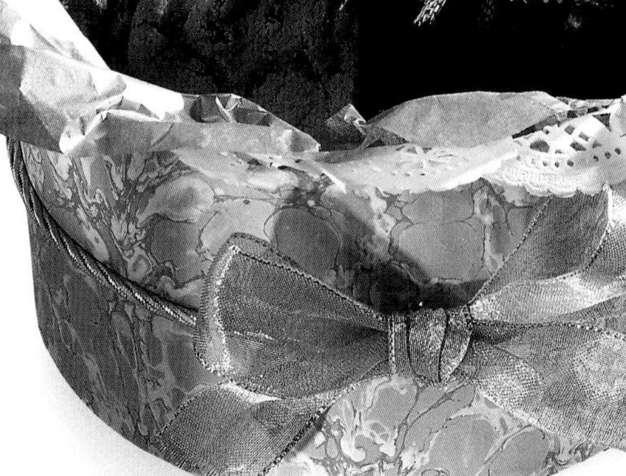

PERFECT FIT
Cover a storage tin and lid with wood-patterned paper, snipping and folding pleats over the edges so they are neat. Adorn with silken braid and gingham tissue paper.

BANDANA WRAP
Bandanas are useful and ornamental. Knot around preserves in a basket.

CRACKING SUCCESS
Cellophane crackers are a good way to display jars of preserves, herbs, and spices.

TEA-TIME TREAT
Sit a pot of jam on shredded tissue inside a cup and saucer, then wrap in cellophane and tie with ribbon.

BASKET WORK
Wooden and straw baskets look attractively rustic. Lay preserves on a bed of straw and secure in place with wire.

LABELS & TAGS

Decorate your gifts with a label or a tag, to provide a personal touch to your culinary creations. Commercial labels, tags, and covers are available, but it is easy to design and make your own. A calligraphy pen gives a professional look.

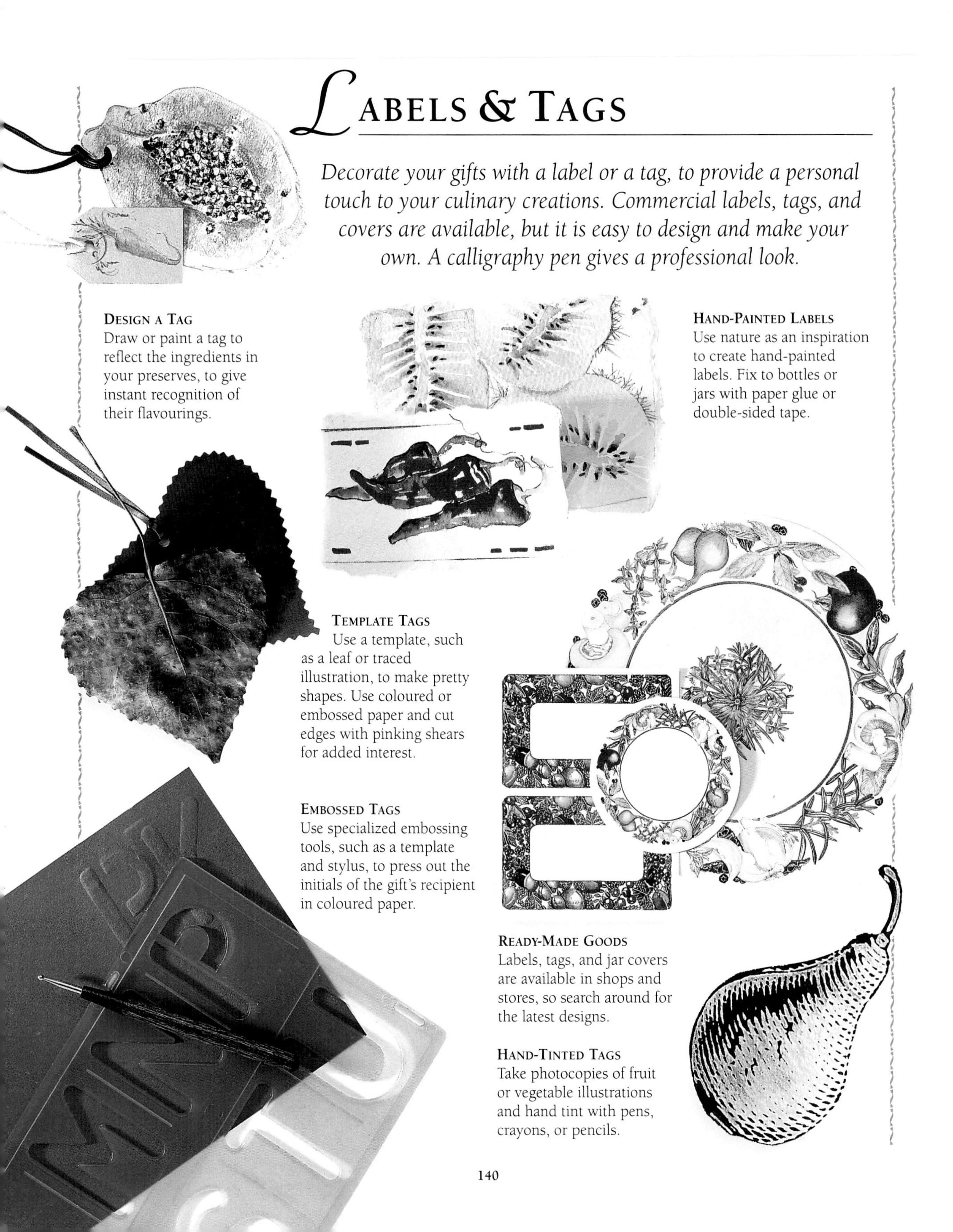

DESIGN A TAG
Draw or paint a tag to reflect the ingredients in your preserves, to give instant recognition of their flavourings.

HAND-PAINTED LABELS
Use nature as an inspiration to create hand-painted labels. Fix to bottles or jars with paper glue or double-sided tape.

TEMPLATE TAGS
Use a template, such as a leaf or traced illustration, to make pretty shapes. Use coloured or embossed paper and cut edges with pinking shears for added interest.

EMBOSSED TAGS
Use specialized embossing tools, such as a template and stylus, to press out the initials of the gift's recipient in coloured paper.

READY-MADE GOODS
Labels, tags, and jar covers are available in shops and stores, so search around for the latest designs.

HAND-TINTED TAGS
Take photocopies of fruit or vegetable illustrations and hand tint with pens, crayons, or pencils.

A

All-American barbecue mix 128
Almond(s)
 apricot and almond chutney 99
 gooseberry and almond conserve 30
 Seville orange marmalade with 33
Apple(s)
 apple and ginger jam 27
 apple and orange jelly 44
 autumn fruit chutney 92
 Bengal chutney 97
 blackberry and apple cordial 69
 blackberry and apple jam 23
 chilli and apple jelly 45
 cider-apple butter 49
 crab apple jelly 46
 cranberry and apple jelly 42-3
 mint and apple jelly 41
 mixed fruit jelly 44
 prune and hazelnut chutney 95
 spiced apples with rosemary 86
Apricot(s)
 apricot and almond chutney 99
 apricot wine with brandy 64
 dried apricot conserve 30
 fresh apricot jam 25
 spiced pickled apricots 89
 tipsy apricots 59
Aromatic garam masala 125
Autumn fruit chutney 92

B

Banana chutney 96
Barbecue mix, all-American 128
Basil
 blueberry-herb vinegar 118
 oil of Italy 112
Beans, see Green beans
Beer, spice mix for mulled 131
Beetroot
 beetroot chutney 97
 pickled diced beetroot 82
Bengal chutney 97
Blackberry(ies)
 blackberry and apple cordial 69
 blackberry and apple jam 23
 blackberry jam 24
 bramble jelly 46
 spiced blackberry vinegar 119
Blackcurrant(s)
 blackcurrant jam 26
 cassis 65
 summer fruits in kirsch 57
Blueberry(ies)
 blueberry-herb vinegar 118
 blueberry jam 22
Blushing strawberry wine 64
Bouquet garni 124, 129
Bouquet of herbs vinegar 121
Bramble jelly 46
Brandy 56
 apricot wine with brandy 64
 brandied carrot jam 21
 cassis 65
 cherry brandy 63
 fruit brandies 62
 mangoes in brandy 58
 peaches in brandy 61
 pears in brandy 59
 vanilla brandy 65
Brining vegetables, for pickles 73
Brown sugar marmalade 33
Butters, fruit, see Fruit butters

C

Cabbage
 end-of-season relish 103
 see also Red cabbage
Capsicum(s)
 apricot and almond chutney 99
 carrot and cucumber relish 103
 end-of-season relish 103
 Jerusalem artichoke relish 102
 Moroccan pickled capsicums 76
 pear and onion chutney 98
 pickled cauliflower with capsicums 79
 spiced capsicum chutney 99
 sweet capsicum pickles 80
 sweetcorn relish 101
Cardamom
 aromatic garam masala 125
 Kashmiri masala 126
Carrot(s)
 Bengal chutney 97
 brandied carrot jam 21
 carrot and cucumber relish 103
 sweet and sour carrots 81
Cashew nuts
 pawpaw chutney 96
 pear and cashew conserve 31
Cassis 65
Cauliflower
 pickled cauliflower with capsicums 79
Chaat masala 126
Cheese
 feta cheese with rosemary in olive oil 116
 goat's cheese with herbs in oil 116
 preserving in oil 113
Cheeses, fruit, see Fruit cheeses
Cherry(ies)
 cherry brandy 63
 pickled cherries 86
Chilli(ies) 45
 carrot and cucumber relish 103
 chaat masala 126
 chilli and apple jelly 45
 fiery chilli oil 109
 hot pepper sherry or rum 65
 perfumed Thai oil 111
 red-hot lemon slices in olive oil 115
Chutneys 90-9
 apricot and almond 99
 autumn fruit 92
 banana 96
 beetroot 97
 Bengal 97
 dark mango 91
 date and orange 93
 lemon and mustard seed 95
 nectarine 92
 orange 96
 pawpaw 96
 pear and onion 98
 pineapple 93
 prune and hazelnut 95
 red tomato 94
 rhubarb 98
 spiced capsicum 99
Cider-apple butter 49
Cider vinegar 72, 117
Cinnamon sugar 130
Citrus fruit(s)
 chopping 38
 marmalades 32
 squeezers 16
Clementines, tipsy 59
Conserves 28-31
 dried apricot 30
 gooseberry and almond 30
 pear and cashew 31
 pear and ginger 31
 plum and orange 31
 prune 29
Containers
 presentation ideas 138-9
 see also Bottles; Jars
Cordials 66-9
 blackberry and apple 69
 kumquat 68
 loganberry and lemon balm 68
 St. Clement's cordial 69
Coriander
 orange and coriander oil 110
 perfumed Thai oil 111
Covering jar lids 136
Covers and collars
 for bottles 135
Crab apple jelly 46
Crackers, presentation idea 139
Cranberry (ies)
 cranberry and apple jelly 42-3
 cranberry butter 49
 cranberry-orange relish 102
Cucumber
 carrot and cucumber relish 103
 dill pickles with garlic 78
Curaçao 56
 tipsy apricots 59
Curds, fruit, see English fruit curds

D

Date(s)
 banana chutney 96
 date and orange chutney 93
 orange chutney 96
 pickled dates 87
Decorating bottles and jars 134-7
Découpage 136
Dill
 dill pickle spices 127
 dill pickles with garlic 78
Dry-brining 73
Dry-frying spices 126

E

Earl Grey tea, spiced
 prunes with 89
Embossed tags 140
End-of-season relish 103
English fruit curds 51-3
 lemon 52-3
 lime 53
 orange 53
 pink grapefruit and lime 53
 tipsy 53
English mixed herbs 129
English mustard 105
Equipment
 jam making 16-17
 pickles 75-6

F

Feta cheese
 with rosemary in olive oil 116
Fine-shred marmalade 36
Five-spice peaches 84-5
Five-spice powder 127
Fruit(s)
 in alcohol 56-61
 chutneys 90-9
 cordials and syrups 66-9
 flavoured vinegars 117
 flavoured wines and spirits 62
 jams 14-15
 jellies 40
 pickles 72-5, 84-9
 preserves 28
 relishes 100-2
 summer fruits
  in kirsch 57
 see also individual types of fruit
Fruit butters 47
 cider-apple 49
 cranberry 49
 peach 49
 strawberry-pear 48
Fruit cheeses 47
 guava 50
 plum 50
 plum and lemon 50
 using moulds 47

G

Garam masala 125
 Indian spice oil 111
Garlic
 crushing 120
 dill pickles with garlic 78
 garlic vinegar 120
Gin 56, 62
 cassis 65
 plum gin 64
 raisins in genever
  with juniper berries 61
Ginger
 apple and ginger jam 27
 grapefruit and ginger marmalade 39
 Japanese pickled ginger 83
 pear and ginger conserve 31
Glühwein, spice mix for 131
Goat's cheese
 with herbs in oil 116
Gooseberry(ies)
 gooseberry and almond conserve 30
 gooseberry jam 27
Grapefruit
 grapefruit and ginger
  marmalade 39
 mandarin marmalade 37
 pink grapefruit and lime curd 53
 pink grapefruit marmalade 37
 three-fruit marmalade 34-5
Grape(s)
 grape jelly 45
 grapes in whisky 60
Greaseproof paper 75, 76
Greek olives in oil 114
Green beans
 mixed vegetable pickles 77
 sweet pickled beans 83
Guava cheese 50

H

Hazelnuts
  prune and hazelnut chutney 95
Herbes de Provence 124, 129
Herb(s)
  all-American barbecue mix 128
  bouquet garni 124, 129
  bouquet of herbs vinegar 121
  English mixed herbs 129
  flavoured mustards 104, 105
  flavoured oils 108
  flavoured vinegars 117
  herbes de Provence 124, 129
  herbs of Provence vinegar 119
  Italian seasoning 129
  jellies 41
  mixed herb oil 111
  Provençal oil 112
  and spice blends 124-9
  whisky liqueur with herbs 65
  see also individual types of herb
Honey
  spiced apples with rosemary 86
Horseradish
  Bengal chutney 97
  horseradish mustard 105
  horseradish relish 102
  pickled horseradish 76
Hungarian oil 110

I

Indian spice oil 111
Infusions
  oils 108
  spice mixes for punches 131
Italian seasoning 129

J

Jams 14-27
  apple and ginger 27
  blackberry and apple 23
  blackcurrant 26
  blueberry 22
  brandied carrot 21
  crushed strawberry 21
  equipment 16-17, 26
  fresh apricot 25
  gooseberry 27
  ingredients 14-15
  loganberry 24
  nectarine 24
  peach 20
  pear and plum 20
  pineapple 23
  plum 26
  plum and walnut 18-19
  raspberry 24
  raspberry and redcurrant 22
  rhubarb and strawberry 25
  strawberry, crushed 21
  strawberry, whole 21
  testing for a set 15
Japanese pickled ginger 83
Jars 10
  decorating 136-7
  for fruits in alcohol 56
  for jam 17
  for pickles 75-6
  sterilizing 11
  see also Sealing and storage
Jellies 40-6
  apple and orange 44
  bramble 46
  chilli and apple 45
  crab apple 46
  cranberry and apple 42-3
  grape 45
  herb 41
  mint and apple 41
  mixed fruit 44
  red fruit with port 41
  testing for a set 46
Jelly bags 40, 41
Jerusalem artichokes
  Jerusalem artichoke relish 102
  pickled Jerusalem artichokes 78
Juniper berries, raisins in genever with 61

K

Kashmiri masala 126
Kirsch 56
  pineapples in kirsch 61
  summer fruits in kirsch 57
Kiwi fruit
  kiwi-passion fruit syrup 69
  peppered kiwi fruit 88
Knives 16
Kosher pickles
  see dill pickles with garlic 78
Kumquat cordial 68

L

Labels 140
Leaves, covering lids with 136
Lemon
  lemon curd 52-3
  lemon and lime marmalade 37
  lemon marmalade 39
  lemon and mustard seed chutney 95
  lemon sugar 130
  plum and lemon cheese 50
  red-hot lemon slices in olive oil 115
  St. Clement's cordial 69
  three-fruit marmalade 34-5
Lemon balm
  loganberry and lemon balm cordial 68
Lemon grass
  perfumed Thai oil 111
Lime(s)
  lemon and lime marmalade 37
  lime curd 53
  pineapple and lime syrup 67
  pink grapefruit and lime curd 53
Liqueurs
  liqueur marmalade 33
  tipsy curd 53
  whisky liqueur with herbs 65
  see also Spirits, flavoured
Loganberry(ies)
  loganberry and lemon balm cordial 68
  loganberry jam 24

M

Malt vinegar 72, 117
Mandarins
  marmalade 36
  tipsy 59
Mango(es)
  dark mango chutney 91
  mangoes in brandy 58
Marmalades 32-9
  brown sugar 33
  dark chunky 33
  fine-shred 36
  grapefruit and ginger 39
  lemon 39
  lemon and lime 37
  liqueur 33
  mandarin 36
  pink grapefruit 37
  quick chunky 38
  Seville orange 33
  Seville orange with almonds 33
  sweet orange 38
  testing for a set 35, 38
  three-fruit 34-5
Measuring equipment 16
Melon
  melon and orange preserve 31
  peach and melon preserve 29
Mint and apple jelly 41
Mixed spices 130
Moroccan pickled capsicums 76
Mould
  on jams 15
  on pickles 74
  on preserves in oil 113
Mulled beer, spice mix for 131
Mushrooms
  mushrooms in oil 115
  tangy pickled mushrooms 82
Muslin 17
  spices in 75, 131
Mustards 104-5
  English 105
  horseradish 105
  lemon and mustard seed chutney 95
  tarragon 105
  whole-grain 105

N

Nectarine(s)
  nectarine chutney 92
  nectarine jam 24

O

Oils
  feta cheese with rosemary in olive oil 116
  fiery chilli oil 109
  flavoured oils 108-12
  goat's cheese with herbs in 116
  Greek olives in 114
  Hungarian oil 110
  Indian spice oil 111
  mixed herb oil 111
  mushrooms in 115
  oil of Italy 112
  orange and coriander oil 110
  perfumed Thai oil 111
  preserves in oil 113-16
  Provençal oil 112
  Provençal olives in aromatic oil 116
  red-hot lemon slices in olive oil 115
  rosemary oil 112
  sun-dried tomatoes in olive oil 115
  sweet paprika oil 110
Olive oil 108, 113
Olives
  Greek olives in oil 114
  Provençal olives in aromatic oil 116
Onion(s)
  pear and onion chutney 98
  pickled onions 74, 76
  spiced capsicum chutney 99
Orange(s) 32
  apple and orange jelly 44
  cranberry-orange relish 102
  date and orange chutney 93
  drying zest 110
  fine-shred marmalade 36
  kumquat cordial 68
  melon and orange preserve 31
  mixed fruit jelly 44
  orange chutney 96
  orange and coriander oil 110
  orange curd 53
  orange-scented vinegar 119
  orange spice mix 131
  orange sugar 130
  plum and orange conserve 31
  quick chunky marmalade 38
  St. Clement's cordial 69
  Seville orange marmalade 33
  spiced oranges 87
  sweet orange marmalade 38
  three-fruit marmalade 34-5

P

Painting
  bottles 135
  labels 140
Paprika oil, sweet 110
Passion fruit
  kiwi-passion fruit syrup 69
Pawpaw(s)
  pawpaw chutney 96
  pawpaws in rum 60
Peach(es)
  five-spice peaches 84-5
  peach butter 49
  peach jam 20
  peach and melon preserve 29
  peaches in brandy 61
Pear(s)
  autumn fruit chutney 92
  discoloration, to prevent 59
  pear and cashew conserve 31
  pear and onion chutney 98
  pear and plum jam 20
  pears in brandy 59
  pears in red wine 60
  pears in vodka 58
  spiced pickled pears 89
  strawberry-pear butter 48
Pectin 14, 15, 20, 32
Peel, citrus fruits 32
Peppercorns
  peppered kiwi fruit 88
Perfumed Thai oil 111
Pickles 72-89
  baby vegetables 78
  cauliflower with capsicums 79
  cherries 86
  zucchini 79
  dates 87
  diced beetroot 82
  dill with garlic 78
  equipment 75-6
  five-spice peaches 84-5
  horseradish 76

ingredients 72-3
Japanese pickled ginger 83
Jerusalem artichokes 78
mixed vegetable 77
Moroccan pickled capsicums 76
onions 74, 76
peppered kiwi fruit 88
preparation 73
problems 74
red cabbage 81
spiced apples with rosemary 86
spiced oranges 87
spiced pickled apricots 89
spiced pickled pears 89
spiced prunes with Earl Grey tea 89
sweet capsicums 80
sweet pickled beans 83
sweet and sour carrots 81
tangy pickled mushrooms 82
walnuts 87
Pickling spice 126
Pickling vinegar, spiced 120
Pineapple(s)
pineapple chutney 93
pineapple jam 23
pineapple and lime syrup 67
pineapples in kirsch 61
Pink grapefruit
pink grapefruit and lime curd 53
pink grapefruit marmalade 37
Plum(s)
autumn fruit chutney 92
pear and plum jam 20
plum cheese 60
plum gin 64
plum jam 26
plum and lemon cheese 50
plum and orange conserve 31
plum and walnut jam 18-19
plums in rum 58
Port
prunes in port 59
red fruit jelly with port 41
Presentation ideas 138-9
Preserves 28-31
melon and orange 31
in oil 113-16
peach and melon 29
strawberry 30
Preserving pans 17, 75
Provençal oil 112
Provençal olives in aromatic oil 116
Prune(s)
prune conserve 29
prune and hazelnut chutney 95
prunes in port 59
spiced prunes with Earl Grey tea 89
Pudding spice 130
Punches, infusing spices 131

Q

Quatre épices 127

R

Raisins in genever with juniper berries 61
Raspberry (ies)
raspberry jam 24
raspberry and redcurrant jam 22
red fruit jelly with port 41
rosy raspberry vinegar 121
Red cabbage
pickled 81
Red fruit jelly
with port 41
Redcurrant(s)
raspberry and redcurrant
jam 22
red fruit jelly
with port 41
Relishes 100-3
carrot and cucumber 103
cranberry-orange 102
end-of-season 103
horseradish 102
Jerusalem artichoke 102
sweetcorn 101
Rhubarb
rhubarb chutney 98
rhubarb and strawberry
jam 25
Rosemary
feta cheese with rosemary
in olive oil 116
rosemary and allspice vinegar 120
rosemary oil 112
spiced apples with rosemary 86
Rum 56
hot pepper rum 65
pawpaws in rum 60
plums in rum 58
tipsy apricots 59
Rumtopf 58

S

Sage
oil of Italy 112
St. Clement's cordial 69
Salt
pickles 72, 73
seasoning salt 128
tarragon salt 128
Sealing and storage
chutneys 90
English fruit curds 50
flavoured oils 108
flavoured vinegars 117
flavoured wines and spirits 62
fruit butters and cheeses 47
fruit cordials and syrups 66
fruits in alcohol 56
herb and spice blends 124
jams 15
jellies 40
marmalades 32
mustards 104
pickles 73
preserves and conserves 28
preserves in oil 113
relishes 100
Seasonings
Italian seasoning 129
seasoning salt 128
Seven Seas spice mix 128
Seville orange marmalade 33
with almonds 33
Seville oranges 32
Sherry, hot pepper 65
Sherry vinegar 117
Sieves 17
Skimming marmalades 38
Spice(s)
aromatic garam masala 125
bags 138
blends 124-31
chaat masala 126
chutneys 90-9
dill pickle spices 127
dry-frying 126
five-spice powder 127
flavoured mustards 104
flavoured oils 108
flavoured vinegars 117
fruits in alcohol 56
Indian spice oil 111
Kashmiri masala 126
mixed spices 130
orange spice mix 131
pickles 72, 75
pickling spice 126
Provençal oil 112
quatre épices 127
seasoning salt 128
Seven Seas spice mix 128
spice mix for *Glühwein* 131
spice mix for mulled beer 131
see also individual types of spice
Spirits, flavoured 62-5
Spoons 16-17
Stencils
decorating bottles with 134
Sterilizing bottles and jars 11
Stones
kernels 14-15
removing 24
Storage, see Sealing and
storage
Strawberry(ies)
blushing strawberry wine 64
crushed strawberry jam 21
mixed fruit jelly 44
rhubarb and strawberry jam 25
strawberry-pear butter 48
strawberry preserve 30
strawberry syrup 68
whole strawberry jam 21
Sugar
chutneys 90
cinnamon sugar 130
English fruit curds 51
jams 14, 15
fruit butters and cheeses 47
fruit cordials and syrups 66
fruits in alcohol 56
jellies 40
lemon sugar 130
orange sugar 130
pickles 72
thermometers 15, 16
vanilla sugar 130
warming 20, 36, 44
Summer fruits
in kirsch 57
Sun-dried tomatoes
in olive oil 115
Sweet and sour carrots 81
Sweetcorn relish 101
Syrups 66-9
kiwi-passion fruit 69
pineapple and lime 67
strawberry 68

T

Tags 140
Tarragon
tarragon mustard 105
tarragon salt 128
tarragon vinegar 121
Tea, Earl Grey, spiced
prunes with 89
Thai oil, perfumed 111
Thermometers 15, 16
Three-fruit marmalade 34-5
Tipsy apricots 59
Tipsy curd 53
Tomato(es)
end-of-season relish 103
horseradish relish 102
pear and onion chutney 98
red tomato chutney 94
spiced capsicum chutney 99
sun-dried tomatoes
in olive oil 115

V

Vanilla
vanilla brandy 65
vanilla sugar 130
Vegetable(s)
chutneys 90-9
mixed vegetable pickles 77
pickled baby vegetables 78
pickles 72-83
preserves in oil 113
relishes 100-3
see also individual types of
vegetable
Vinegars
blueberry-herb 118
bouquet of herbs 121
flavoured 117-21
garlic 120
herbs of Provence 119
orange-scented 119
for pickles 72, 74
rosemary and allspice 120
rosy raspberry 121
spiced blackberry 119
spiced pickling 120
tarragon vinegar 121
Vodka, pears in 58

W

Walnut(s)
pickled walnuts 87
plum and walnut jam 18-19
Warming sugar 20, 36, 44
Wax seals 137
Whisky 56
grapes in whisky 60
whisky liqueur with herbs 65
Wine 56
apricot wine with brandy 64
blushing strawberry wine 64
cassis 65
flavoured 62-5
pears in red wine 60
Wine vinegar 72, 117

Z

Zest, drying 110
Zucchini pickles 79

Acknowledgments

Photographers David Murray
Jules Selmes
Photographer's Assistant Steven Head
Home Economist Sarah Lowman

Typesetting Debbie Lelliott
Linda Parker
Debbie Rhodes

Production Consultant Lorraine Baird
Text film by Disc to Print (UK) Limited

Carroll & Brown Ltd would like to thank Stephen Poole, Mary Denning, Pat Baines, and Audrey Fox for supplying produce from their gardens for photography and Carolyn Chapman for her help with the Finishing Touches chapter. A selection of decorative ribbons was supplied by Panda Ribbons. Labels on pages 132 and 137 are copyright of Kate Weese, California.

Notes

- Preserving is not without dangers. Cleanliness, equipment, timing, acidity, and a wide variety of other factors are critical to getting results that are safe to eat. The foods shown in this book were decanted and photographed immediately after preparation; they are not all in airtight containers. This book gives recipes and guidelines only. If you wish to store foods for longer periods, you must follow specific instructions for methods of preserving. Dorling Kindersley assumes no responsibility for the preserving of foods described in this book.
- Over a period of prolonged storage, the colours of some of the preserves may fade.